Practical Oilfield Metallurgy and Corrosion

PRACTICAL OILFIELD METALLURGY AND CORROSION

SECOND EDITION

By Bruce D. Craig

PennWell Books
Pennwell Publishing Company
Tulsa, Oklahoma

PennWell Publishing Company
1421 South Sheridan/P.O. Box 1260
Tulsa, Oklahoma 74101

Craig, Bruce D.
Practical oilfield metallurgy and corrosion/by Bruce D. Craig. — 2nd ed.
p. cm.
Rev. ed. of: Practical oil-field metallurgy. 1984.
Includes bibliographical references and index.
ISBN 0-87814-388-2
1. Oil field — Equipment and supplies. 2. Metallurgy. 3. Corrosion and anti-corrosivess. I. Craig, Bruce D. Practical oil-field metallurgy. II. Title.
TN871.5.C7 1992
622'.3382—dc20 92-33961
CIP

Printed in the United States of America

TO

My Wife

and

My Father and **Mother**

Three very
important people
in my life

Contents

Preface

As indicated in the first edition, this book continues to be the only book written specifically on the subject of metallurgy in petroleum production. It is by no means comprehensive in its approach either to metallurgy or to petroleum production. In fact, a reader well versed in certain specialty areas of the industry will notice gaps that are unavoidable because of either the author's ignorance in the area or a lack of available information in the literature. The latter is quite frequent because many tools and processes are proprietary and therefore information is not readily available.

The purpose for writing this book was twofold. First, there are many nonmetallurgists in the petroleum industry who are responsible for materials selection and failure analysis, yet no single source is available for reference on the subject. Second, for the practicing metallurgist in the petroleum industry, no reference book contains many of the often-used tables, charts, graphs, etc., that are helpful in day-to-day decision making on materials. For these two groups, this book was written.

Oilfield jargon for equipment and operations is used as often as possible to acquaint the reader with the proper nomenclature and to provide the experienced oilfield engineer with a familiar reference point. Where possible, actual examples from oilfield materials and failures are used to further assist the reader in understanding this discipline.

In the past, knowledge of oilfield metallurgy has been acquired by a combination of learning from experienced individuals in the field and by trial and error. This method of education, while effective, is quite time consuming and costly. The author hopes this book will reduce some of the front-end time of acquainting oneself with the fundamentals prior to becoming an effective materials engineer. This is particularly important at this stage in the history of the petroleum industry because the capabilities of alloys are being challenged in severe environments, and mistakes in application can be of tremendous economical consequence — not to mention the potential for loss of life.

Chapter One of this book is designed to give the reader a basic understanding of metallurgy and to aid in the discussion of metallurgy in subsequent chapters. Chapter Two introduces some of the major aspects of corrosion as they relate to the petroleum industry. Corrosion is an inseparable part of oilfield metallurgy, and the two will be linked throughout the book. While in some ways this may appear to be a corrosion book, it is not. Only metallurgical alternatives to problems are described, whereas other alternatives such as inhibitors, coatings, etc., exist that are not discussed but do enter in corrosion-control planning. Chapters Three, Four, Five, and Six describe the applications and behavior of alloys in various functions in the industry. Certain aspects of oilfield metallurgy have been intentionally deleted for clarity. One of these is welding, which, although an important part of metallurgy, would tend to add to the complexity of this book beyond any potential benefit. Finally, Chapter Seven is a brief introduction to failure analysis, which is a very important part of oilfield metallurgy.

As in the first edition, the author would like to acknowledge those whose work is referenced in this book and the many friends in the industry who are not named but who have generously contributed to his education and understanding, not only of metallurgy but also of the field of petroleum engineering. Special mention is due R.E. Lewis of Texaco and H.M. Shaffer of Marathon Oil Company for their years of instruction and discussion on corrosion in the oil industry and, of course, for their valuable friendship. The author also acknowledges American Petroleum Institute, 1220 L Street Northwest, DC 20005 for the use of their specification tables throughout the book. The second edition represents the influence of a broader circle of friends and colleagues in the industry. The author especially acknowledges one of the most important of these friends and colleagues; Ted Bruno of Metallurgical Consultants Inc. and of course, the other characters who are MCI are thanked. Finally, appreciation is extended to Nina Irwin for bringing order out of chaos by typing the manuscript.

1

BASIC METALLURGY

All metals except mercury are solids at room temperature. These solids consist of atoms that form a regular array, or pattern, called a *crystal lattice*. The nuclei of these atoms describe the lattice points and the tightly bound electrons remain close to these atom centers. *Valence electrons*, or the outermost, loosely bound electrons, move easily throughout the lattice. These valence electrons are the primary reason for some of the superior properties of metals as compared to nonmetallic solids. Electrical and thermal conductivity, magnetism, and luster are the results of the behavior of these valence electrons.

Structure of Metals

Energy considerations as well as the nature of metal bonding produce several different types of lattice structures, three of which are most frequently observed in metals. These three structures, or unit cells as they are most commonly referred to, are *face-centered cubic* (fcc), *body-centered cubic* (bcc), and *hexagonal close packed* (hcp). The first structure (see Fig. 1–1) is portrayed as a group of spheres packed closely together. Each sphere represents an atom, which is an idealized but simple way to describe atom packing. The unit cell is the smallest increment to which the periodic structure of metal can be reduced.

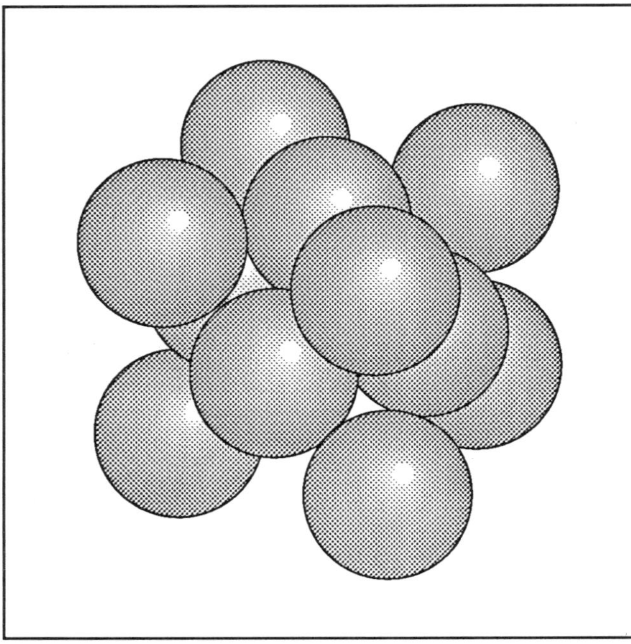

Figure 1–1 Hard sphere model of face-centered cubic atom packing.

Examples of metals with fcc structures are copper, nickel, aluminum, gold, and silver. Figure 1–2 shows the stacking arrangement of a bcc unit cell. Note the greater amount of space present in this structure versus the fcc. Examples of metals with this type of unit cell are: iron, molybdenum, chromium, and vanadium. Figure 1–3 depicts the third unit cell, hcp. Magnesium, zinc, cobalt, and titanium are hcp metals. Some metals display different crystal structures as a function of temperature. For example, the high-temperature crystal structure of titanium is bcc while the room-temperature structure is hcp. These different crystal structures in one metal are called *allotropes.*

Mechanical properties such as strength, ductility, and toughness of metals and alloys are strongly dependent on their crystal structure. Hexagonal close-packed metals commonly have less strength than the fcc and bcc metals. Because of the packing arrangement, their ductility is more dependent on direction than fcc and bcc metals. This often precludes their use as structural alloys in many applications, especially in the oilfield.

Metals do not consist of one large crystal with uninterrupted periodicity of unit cells. Rather, as metals cool from the liquid state, nuclei of solid particles form in the liquid. These nuclei continue to grow during cooling until the solid crystals begin to impinge on each other. Two grains having planes of random

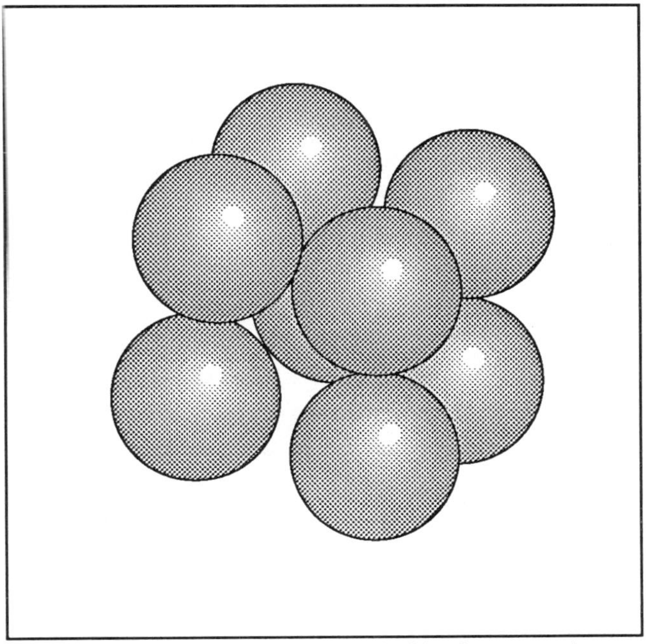

Figure 1–2 Hard sphere model of body-centered cubic atom packing.

orientation to each other will not match at this interface, therefore, some atomic accommodation must result. This accommodation or transition at the junction of two or more grains produces an interface, or *grain boundary*. Since atoms are not in their proper place in this boundary and the periodic nature of the grain has been disrupted, the grain boundary has a higher energy than the surrounding matrix. Because of this different atomic structure and boundary energy, grain boundaries have properties that differ from the grains.

A polycrystalline metal has more strength than a single crystal because of grain boundaries. In fact, as grain size is reduced (which increases the total number of grain boundaries), the strength and ductility of a metal is increased, as shown in Figure 1–4. The yield strength of two common low alloy steels, 8650 and 4340, is a function of the grain size. Fine grain size improves not only strength but also toughness. Toughness is the ability of a material to resist fracture.

Many other ramifications of grain boundaries are not appropriate to elucidate here, but one of significance is *chemical resistance*. Since grain boundaries have higher energy than the grains, they are attacked more rapidly in a corrosive environment. This behavior allows the study of metal structure by optical microscopy and is called *metallography*. Figure 1–5 shows an example of steel that

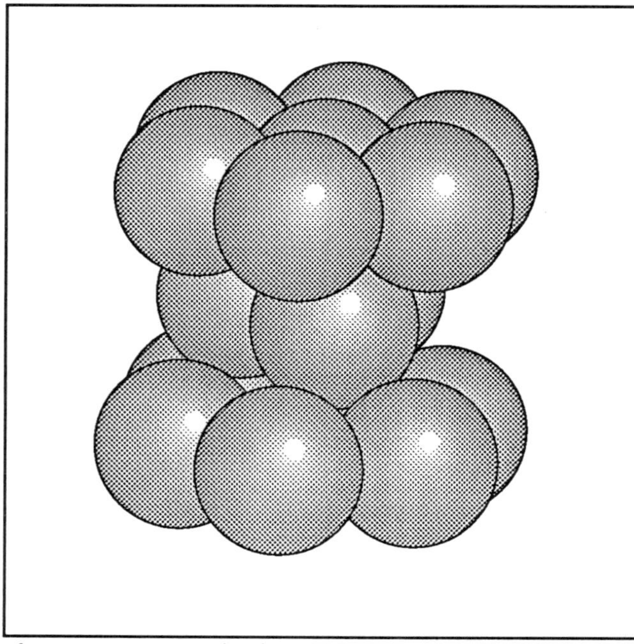

Figure 1–3 Hard sphere model of hexagonal close-packed structure.

has been attacked (etched) by a solution of nitric acid and methanol to resolve the grains and grain boundaries. The etching solution is commonly known as nital.

Pure metals, even in polycrystalline form, do not possess the strength necessary for modern engineering applications. Introducing other elements into a pure metal can provide necessary additional strength. However, the addition of other elements is not restricted solely to the attainment of higher strength. Other elements can provide greater corrosion resistance, increased ductility, enhanced toughness at low temperature, stabilization of the structure against deformation at high temperatures, and a multitude of other properties.

This combination of two or more elements, when at least one is a metal, is called an *alloy*. The mixture of these elements, defined as a solid solution, can occur in two ways. In one way (see Fig. 1–6), the elements substitute themselves on the atom sites of the parent metal. In the other (see Fig. 1–7), the elements fit in the interstitial holes between the parent metal atoms. The first situation is referred to as *substitutional solid solution* while the second is *interstitial solid solution*. The preference for one type of solid solution versus the other is a function of the sizes of the intervening solute atoms compared to the size of the solvent atoms. The base metal is referred to as the solvent and the

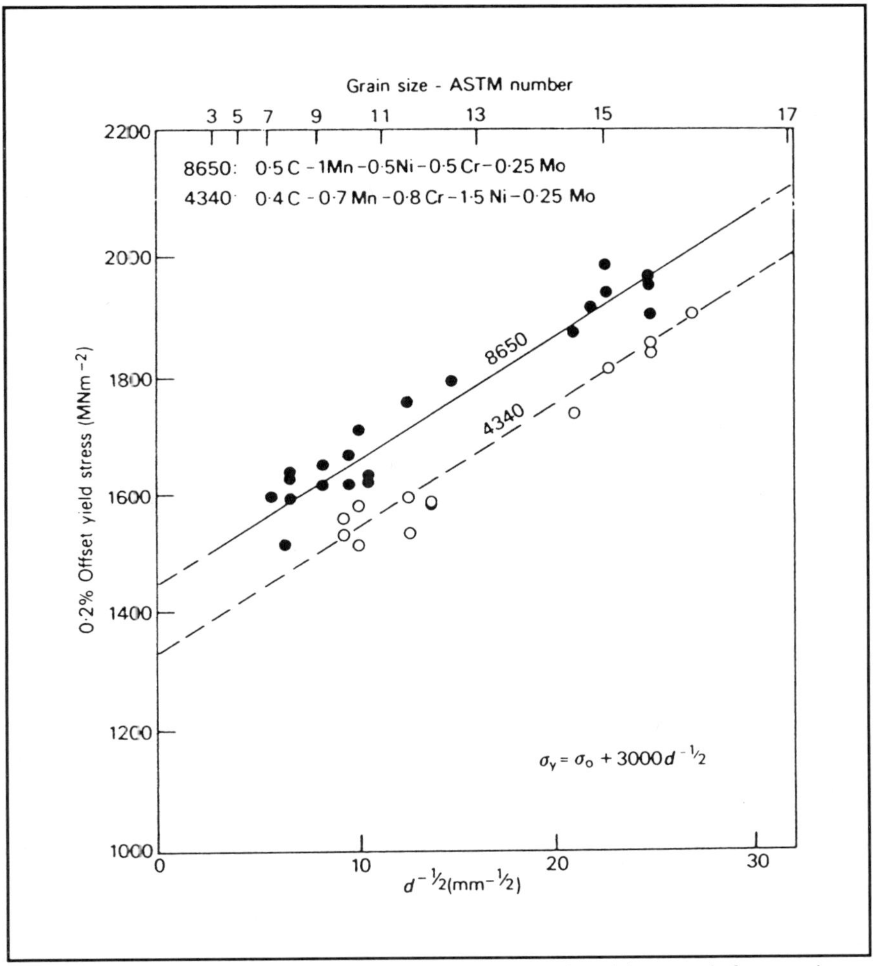

Figure 1–4 Relationship between grain size and yield strength for two low alloy steels (courtesy ASM International).

addition of another element is the solute. If the solute atom is small, such as carbon, nitrogen, or hydrogen, circumstances will be energetically more favorable for the atom to sit in an interstitial site than in a substitutional one. The opposite is true for large atoms such as chromium, nickel, and manganese. Thus alloys such as steel that contain carbon in iron represent an interstitial solid solution, while monel which is copper in nickel is an example of a substitutional solid solution.

The concentration of alloying elements and the type of solid solution contribute to the properties of an alloy much as described for atom packing and grain

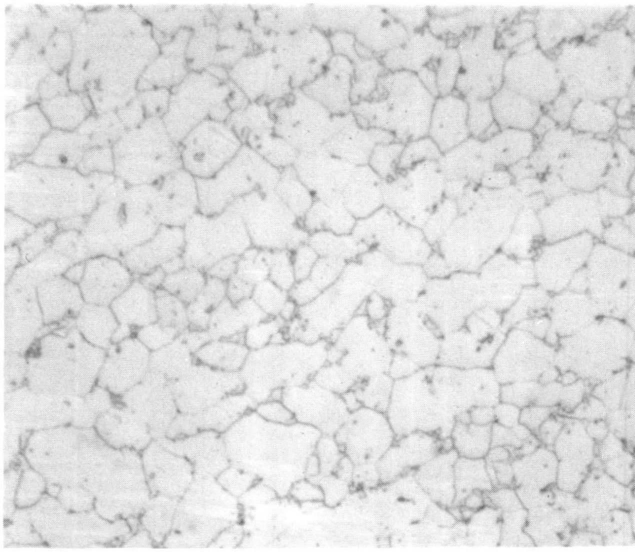

Figure 1–5 Low-carbon (0.12 wt%) steel etched with nital. Magnified 730 times showing grains and grain boundaries.

size. Interstitial solid solutions such as carbon in iron provide a higher degree of strengthening than substitutional solid solutions.

Considering all of these methods for increasing the strength of metals, how can their effects be measured? The next section discusses various test methods employed to determine the mechanical properties of an alloy.

Mechanical Properties of Metal

The most common test for quantitative information on the mechanical properties of a metal is the *tensile test*. In simple terms, for this test a metal is pulled apart at a constant rate. While this test has many sample geometries, one of the standard samples is a round bar of unspecified length that has a reduced section greater than 2 in. in length and is 0.500 in. in diameter. An example of such a bar is shown in Figure 1–8. As the sample is pulled in a tensile machine, the applied load divided by the original cross-sectional area at the center of the sample gives the *stress*. At the same time, a 2-in. length called the gauge length, which is marked on the reduced section of the bar, is used to measure the change in length of the sample in comparison to its original length — the *strain*.

Figure 1–9 shows a common engineering stress-strain curve that results

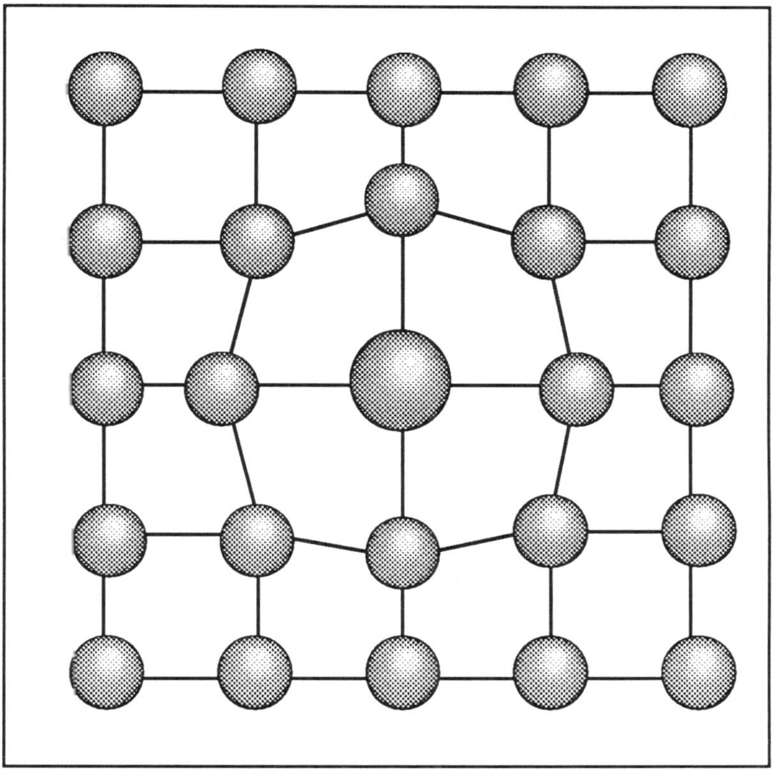

Figure 1–6 Substitutional solid solution of a larger atom substituted on a lattice site of smaller atom base metal.

when the stress is calculated at each point by dividing the load applied by the original cross-sectional area:

$$\sigma = \frac{P}{A}$$

Where:

σ = stress, psi
P = applied load, lb
A = original cross-sectional area, in.^{2}

This is defined as the *engineering stress*. The *engineering strain* (ε) is likewise defined as the total elongation minus the original gauge length divided by the original gauge length:

$$\varepsilon = \frac{\Delta l}{l}$$

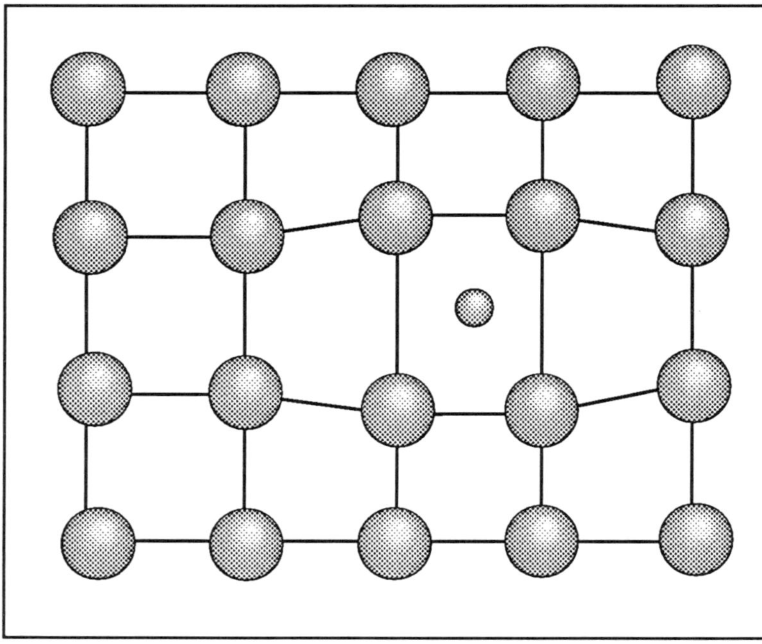

Figure 1–7 Interstitial solid solution of small atom between larger atoms of the base metal.

Where:

Δl = change in length, in.

l = original length, in.

In many tests, the standard gauge length is 2 in.

Two important points on the curve are denoted in Figure 1–9. The first, point A, represents the termination of a linear relationship between stress and strain. This point is referred to as the proportional limit and represents the limit of elastic behavior of the metal. Stresses or strains applied below this limit, when relaxed, will not cause a permanent set or deformation in the metal. From Hooke's law, the relationship between stress and strain is given by:

$$\sigma = E\epsilon$$

Where:

σ = stress

E = elastic constant (Young's Modulus)

ϵ = strain

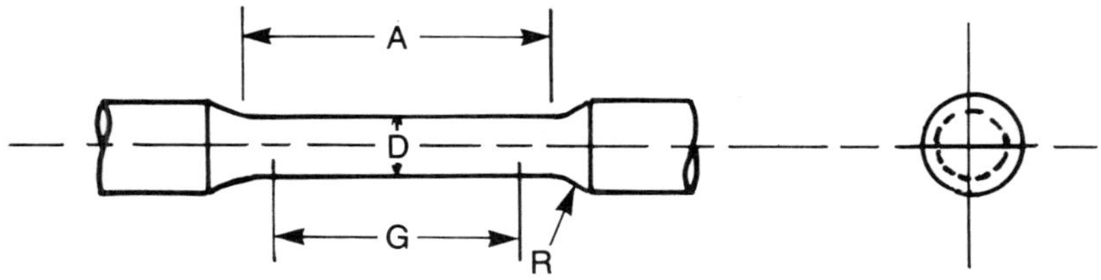

	Standard Specimen	
	in.	mm
	0.500	12.5
G — Gauge length	2.000 ± 0.005	50.0 ± 0.10
D — Diameter	0.500 ± 0.010	12.5 ± 0.25
R — Radius of fillet, min.	3/8	10
A — Length of reduced section min.	2¼	60

Figure 1-8 Standard tensile specimen for determining mechanical properties of a metal (© ASTM, A 370, reprinted with permission).

The units of stress are expressed in pounds per square inch (psi) and strain is unitless since it is length divided by length. Strain is commonly given in percent. The *elastic constant*, termed the modulus of elasticity or Young's Modulus, is also expressed in psi. For steels the Young's Modulus is typically 30 x 10^6 psi.

Once point A is exceeded, the strain becomes plastic and a permanent deformation is produced in the metal. Continuing up the curve, at point B the sample cannot support the increased stress so the cross section begins to reduce itself or neck down. Once necking begins, complete fracture occurs quickly as shown by the change in slope of the curve. The maximum stress attained prior to necking divided by the cross-sectional area is defined as the *ultimate tensile strength*. The amount of necking is directly related to the ductility of a metal. The more necking that occurs, the more ductile it is. A measure of this ductility is the percent reduction in area (%RA). This value is measured from the original cross-sectional area of the tensile specimen and the final cross-sectional area after fracture.

One other measure of ductility is the elongation or stretching of the tensile specimen. If there is significant elongation during the tensile test, the material is quite ductile. Therefore, the change in length of the specimen after testing compared to its original length (most often 2 in. or 50 mm) is reported as a percent elongation.

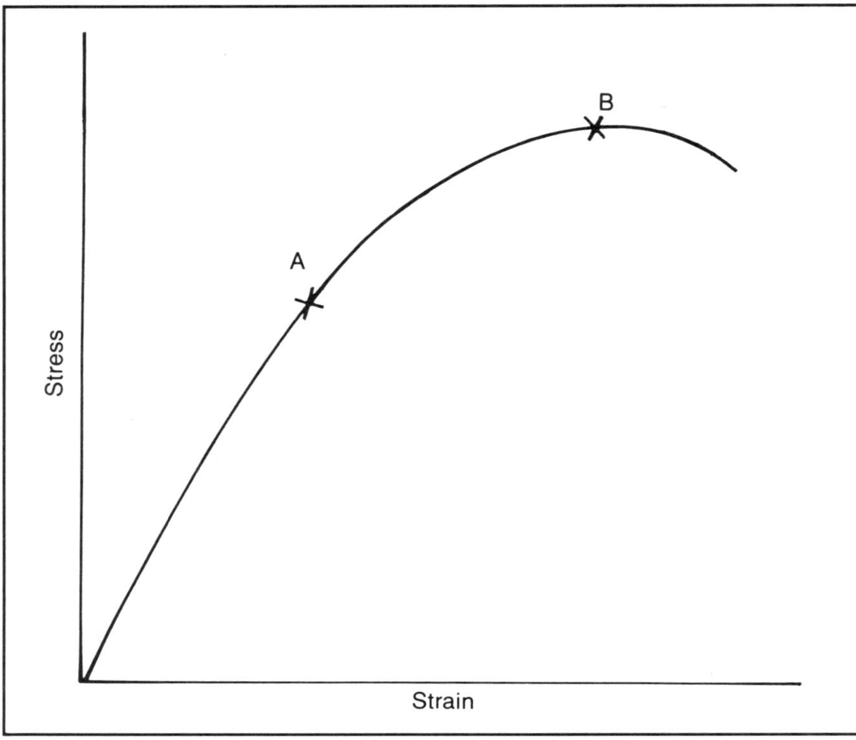

Figure 1–9 Stress-strain curve for a pure metal showing the yield strength (A) and tensile strength (B).

Since the cross section begins to reduce during necking, it is apparent that dividing the stress by the original cross-sectional area will induce a degree of error. In fact, the same holds true for the strain since its length is continuously changing. Therefore, a more accurate means of describing stress and strain, particularly in the plastic region, uses the instantaneous area and length, respectively. This type of calculation gives the true stress and true strain but can also become difficult to measure; thus most design data are presented as engineering stress and strain.

It is apparent from Figure 1–9 that determining the specific point at which the strain moves from elastic to plastic is quite difficult. Therefore, another aid to engineering is the 0.2% *offset yield point.* This commonly used technique requires drawing a line tangent to the elastic portion of the curve then drawing a similar line parallel to the tangent at a strain of 0.2%. The intersection of this line with the stress-strain curve designates the yield strength. The yield strength is an important property of a metal or alloy and is used as a reference for calculating allowable or design stresses of a component or structure.

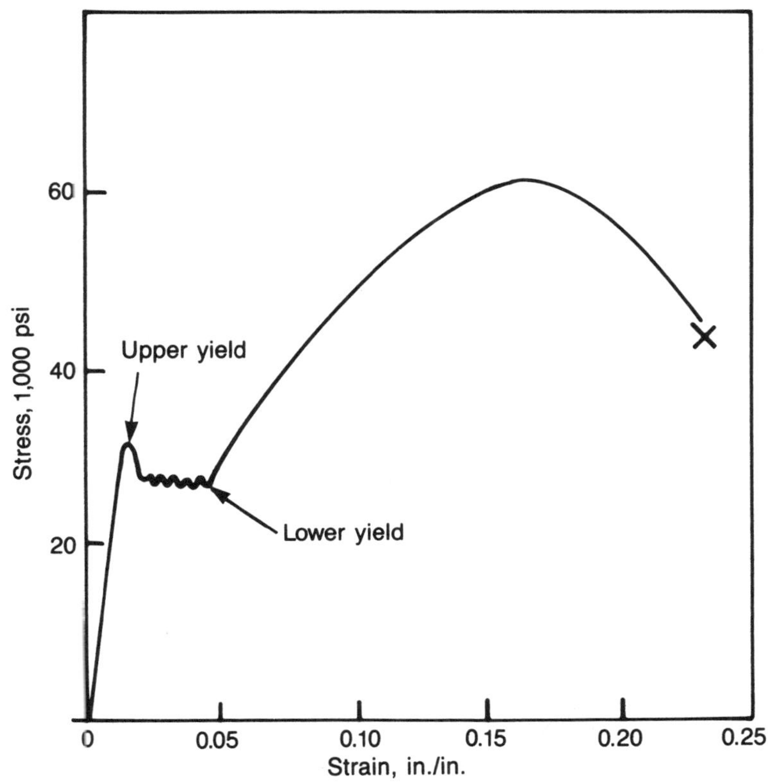

Figure 1–10 Stress-strain curve for AISI 1030 steel showing an upper and lower yield strength.

Of course, the complexity in defining yield point increases in some alloy systems. Figure 1–10 shows a typical curve for a carbon steel with an upper and lower yield point, neither of which corresponds with the 0.2% offset yield point. Steels used in the petroleum industry usually are not complicated by an upper and lower yield; however, the 0.2% offset yield is not the only method of defining the yield strength. Rather, various API tubular specifications require a yield point defined by a 0.5 to 0.7% extension of the gauge length. For most cases, this does not produce a significantly different yield from that designated by the 0.2% offset method, at least for carbon steels. However, for cold-worked nickel alloys the difference can be significant because of the higher strain-hardening rate of these alloys compared to steel. Figure 1–11 presents an example of an API J-55 casing steel stress-strain curve showing the minor difference in yield strength obtained by the two methods.

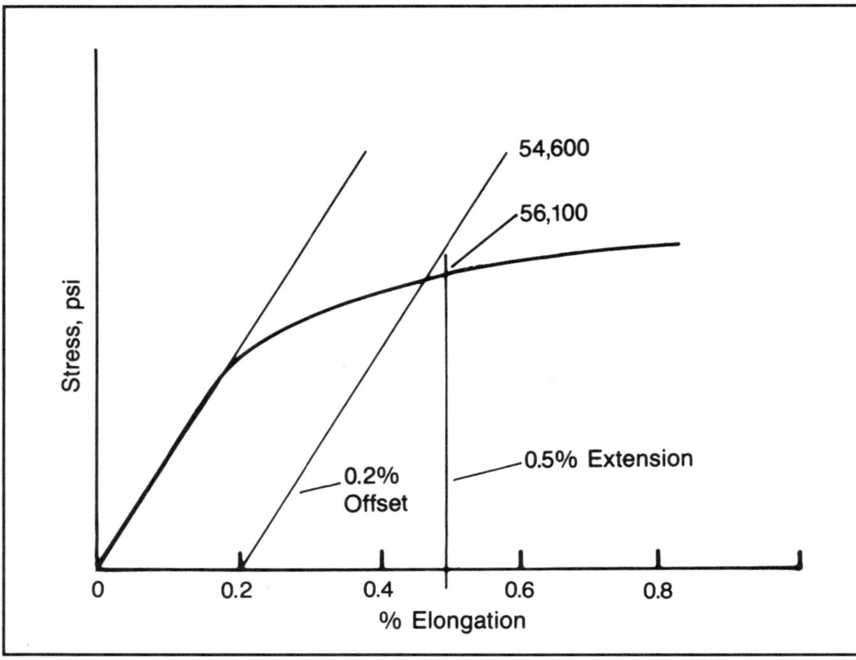

Figure 1–11 Yield strength of J-55 casing determined by 0.2% offset method and 0.5% extension method.

Tensile testing is quite useful in determining the properties of a metal, but it is also a destructive test. In many instances, a nondestructive method of estimating these properties is necessary. The *hardness test* allows an alloy to be tested without destruction and on a smaller scale. Hardness testing requires applying a known load to an indentor and subsequently determining the resistance to this indentation. The higher the strength of the metal, the less deformation will occur under the indentor and the shallower the resulting indentation will be.

Although there are a multitude of hardness tests, only a few are frequently used with metals. These are shown in Figure 1–12. The *Brinell test* is used for field testing large structures or for rapid testing of redundant items (e.g., casing, tool joints, etc.). The impression of the indentor using a 3,000-kg load is 2.25 mm–5.60 mm in diameter for steel.

The *Rockwell hardness test* is used both in field testing and in the shop or laboratory. The various scales are interrelated and have certain advantages and disadvantages over each other. However, the B and C scales are the most common in the oilfield and are overlapping scales. The B scale is used for soft materials with approximate tensile strengths ranging 57,000–115,000 psi, while

Common Hardness Tests

Test	Indentor	Typical Load	Shape of Indentation
Brinell	10 mm sphere (steel or tungsten carbide)	3,000 kg	
Rockwell			
A } C }	Diamond cone	60 kg 150 kg	
B	1/16-in. diameter ball	100 kg	
Vickers	Diamond pyramid	1 kg, 10 kg	
Knoop	Diamond pyramid	1 kg, 10 kg	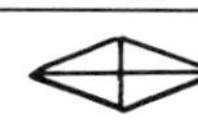

Figure 1–12 Various indentors and types of hardness measurements commonly used in the oil industry.

the C scale ranges from 110,000–over 330,000 psi. These approximate tensile strengths are used only for steel because other alloys have different relationships between hardness and tensile strength.

The approximate ultimate tensile strengths derived from the hardness tests are quite useful, especially in failure analysis when, because of limitations in sample size, obtaining complete tensile data is impossible. From hardness measurements, an approximate ultimate tensile strength can be determined. By using the rule of thumb that for carbon and low-alloy steel the yield strength is approximately 75–90% of the ultimate tensile strength, a yield strength can be approximated.

The two remaining hardness tests are usually referred to as *microhardness tests* since the size of their indentations is on the order of several microns (micrometers), although the ASTM defines microhardness based on loads less than 1 kg. These tests are also most frequently used in the laboratory for analyzing microstructures. The loads shown in Figure 1–12 represent only the more common ones used although a wide range of loads is available for microhardness testing depending on the application. Table 1–1 presents the correspondence between hardness scales and approximate tensile strength.

The decision as to which hardness test to use for a particular application depends on factors such as the number of tests to perform, the sensitivity and accuracy required, and the size and geometry of the sample to be tested.

Table 1-1. Relationship between Common Hardness Scales and Approximate Tensile Strength for Steel.

Rockwell Scale B	Rockwell Scale C	DPH, 10 kg	KHN 500 g and Over	BHN 3,000 kg	Tensile Strength, 1,000 psi approx.
	70	1,076	972		
	69	1,004	946		
	68	942	920		
	67	894	875		
	66	854	870		
	65	820	846		
	64	789	822		
	63	763	799		
	62	739	776		
	61	716	754		
	60	695	732	614	
	59	675	710	600	
	58	655	690	587	
	57	636	670	573	
	56	617	650	560	
	55	598	630	547	301
	54	580	612	534	291
	53	562	594	522	282
	52	545	576	509	273
	51	528	558	496	264
	50	513	542	484	255
	49	498	526	472	246
	48	485	510	460	237
	47	471	495	448	229
	46	458	480	437	221
	45	446	466	426	214
	44	435	452	415	207
	43	424	438	404	200
	42	413	426	393	194
	41	403	414	382	188
	40	393	402	372	182
	39	383	391	362	177
	38	373	380	352	171
	37	363	370	342	166
	36	353	360	332	162
	35	343	351	322	157
	34	334	342	313	153
	33	325	334	305	148
	32	317	326	297	144
	31	309	318	290	140
	30	301	311	283	136
	29	293	304	276	132
	28	285	297	270	129
	27	278	290	265	126
	26	271	284	260	123
	25	264	278	255	120

Rockwell Scale B	Rockwell Scale C	DPH, 10 kg	KHN 500 g and Over	BHN 3,000 kg	Tensile Strength, 1,000 psi approx.
	24	257	272	250	117
100.0	23	251	266	245	115
99.0	22	246	261	240	112
98.5	21	241	256	235	110
97.8	20	240	251	227	108
96.7	(18)	230	236	219	106
95.5	(16)	220	231	210	102
93.9	(14)	213	221	205	98
92.3	(12)	206	211	195	94
90.7	(10)	196	205	187	90
89.0		188	196	179	87
86.8		178	188	179	83
85.0		171	178	163	79
82.9		163	170	156	76
80.8		156	163	149	73
78.7		150	157	143	71
76.4		143	150	137	67
74.0		137	144	131	65
72.0		132	140	126	63
69.8		127	135	121	60
67.6		122	129	116	58
65.7		117	124	111	56

Values in parenthesis are not valid but are given for information only. (After *Metal Progress* and ASM *Metals Handbook*)

Two other properties of metals of consequence in the oilfield are impact and fatigue. The first property deals with the transition in fracture behavior with temperature from ductile to brittle. The second property is the fracture of metals as a result of cyclic loading.

Most body-centered cubic metals and their alloys show a temperature dependence on resistance to crack propagation, whereas fcc and hcp metals do not. Figure 1–13 shows this behavior. Notice that the fcc metals are very resistant to fracture at all temperature while the hcp metals have poor fracture resistance (toughness) at all temperatures. The better resistance of fcc metals to low temperature fracture is the reason for their wide use in cryogenic service. For bcc metals at room temperature and above, the energy necessary to cause rapid unstable crack propagation is quite high as shown by the shelf at the right. At very low temperatures, little energy is required to induce crack propagation. In the middle of these two extremes, a transition region exists in which the fracture changes from a ductile, energy-absorbing fracture to a brittle, low-energy fracture with decreasing temperature.

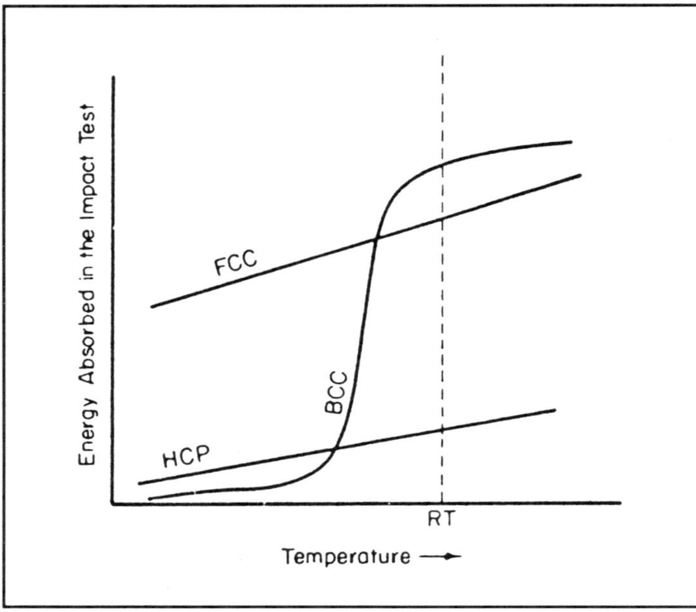

Figure 1–13 Fracture transition behavior as a function of temperature and crystal structure.

The actual temperature at which this transition occurs is difficult to ascertain and open to individual interpretation. Thus, there are many methods to define a transition temperature. Two of the most common are the energy absorbed criterion, frequently 15 ft-lb, and the *fracture-appearance transition temperature* (FATT). Both of these will be explained, but first an explanation of the impact test will elucidate the meaning of these two transition temperature criteria. Figure 1–14 represents a typical ductile-to-brittle transition curve that has been derived by means of Charpy impact testing. The method is described in detail in *ASTM Standard E23*, and the specimen geometry is shown in Figure 1–15. This test is the most common method for evaluating the resistance to brittle fracture of a metal and for determining its transition temperature. With the advance of fracture mechanics in the last 30 years, more sophisticated tests have been devised that give an accurate and quantitative picture of the fracture process. However, the Charpy test remains the fastest and least expensive technique for transition temperature determination.

Returning to the two-transition temperature criteria, energy-absorbed value is easily determined by finding the intersection of this desired value with the curve. The resulting temperature is considered the absorbed-energy transition temperature. For example, if 15 ft-lbs is used then the transition temperature

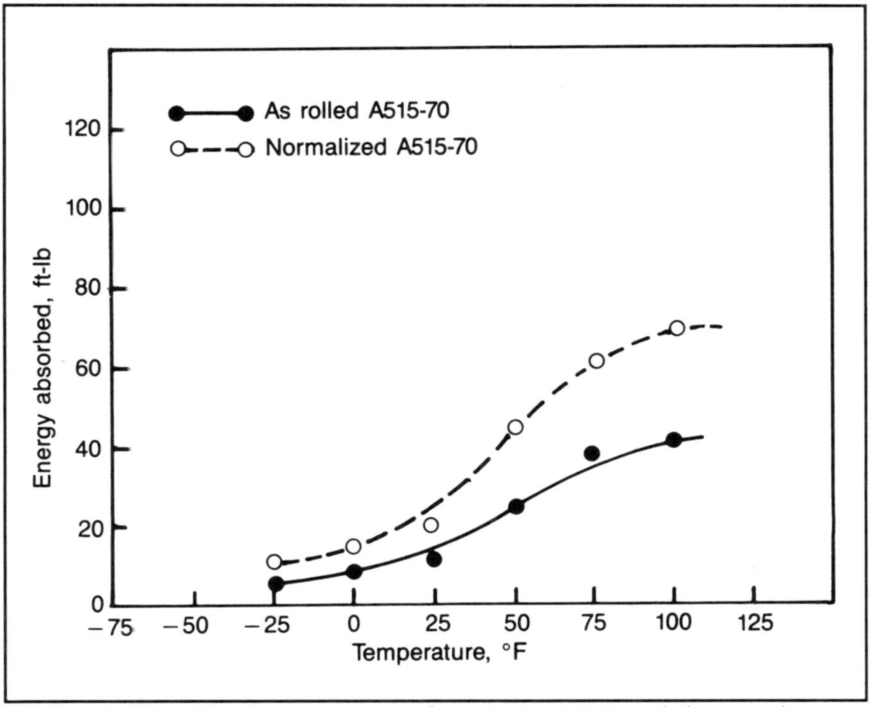

Figure 1–14 Charpy impact curve for ASTM A 515 steel showing the temperature dependence of fracture (After Bravenec).

for normalized A 515 Grade 70 (Fig. 1–14) is approximately 0°F. The FATT is found by plotting the percent of shear observed on the fracture surface versus the temperature. The temperature at which there is 50% shear fracture is often considered the FATT.

Although many parameters affect the transition temperature of an alloy, grain size is probably one of the most important because it can be controlled to provide both higher strength and better toughness. Everything else being equal, the smaller the grain size, the lower the transition temperature and the higher the yield strength. Alloying elements and heat treatment also affect the transition temperature. In steels, a decreased carbon content and a quenched and tempered microstructure generally offer better fracture resistance than a normalized structure. Toughness as a function of carbon content and grain size is shown in Figures 1–16 and 1–17, respectively.

Fatigue is the application of a fluctuating stress, either tensile or compressive. The magnitude of the alternating stress determines the service life of a component. Figure 1–18 shows an idealized fatigue or S-N (stress versus number of cycles) curve. The S-N curve depicts a flat region at the right that

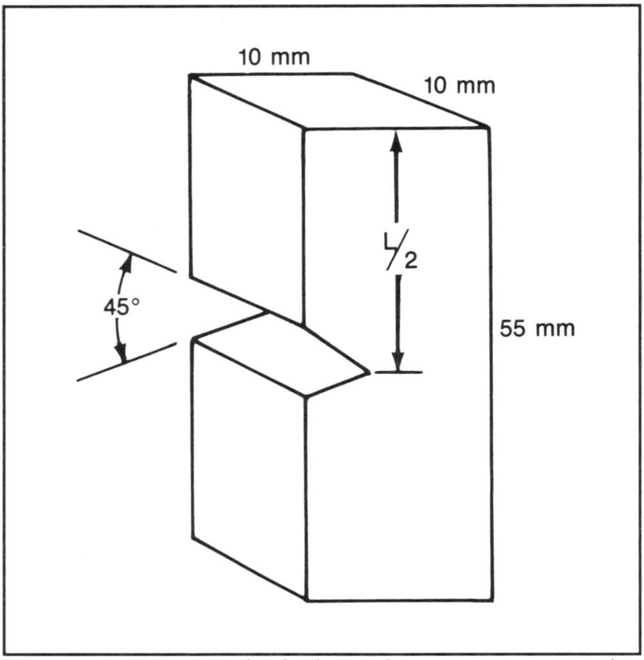

Figure 1-15 Standard Charpy bar geometry according to ASTM Standard E23.

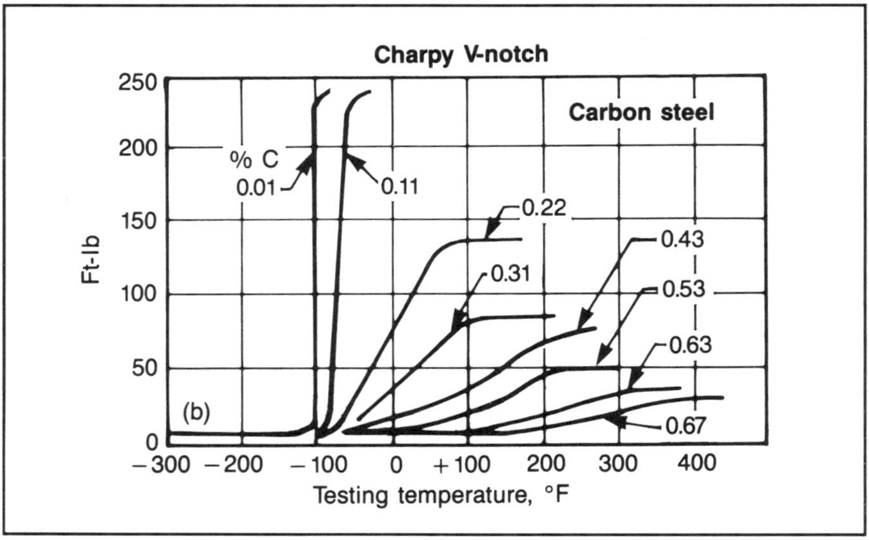

Figure 1-16 Effect of carbon content on notch toughness of mild steel (After ASM *Metals Handbook*).

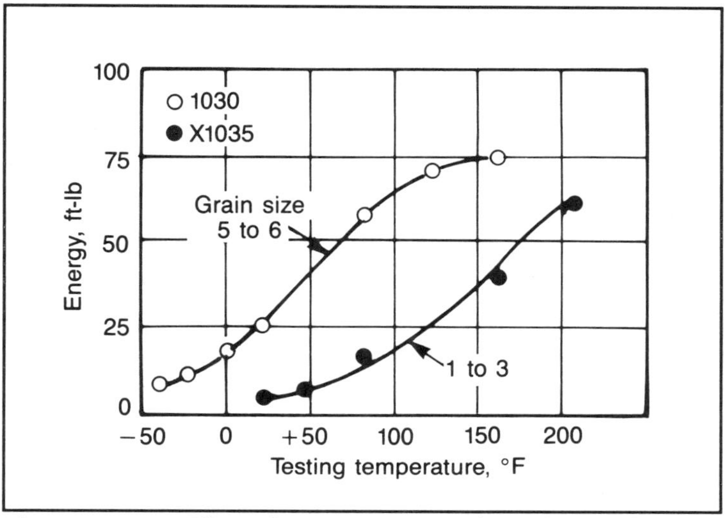

Figure 1–17 Effect of grain size on notch toughness of AISI 1030 and 1035 steel. Larger numbers indicate smaller grain size (After *Metals Handbook*).

represents a threshold or endurance limit. Below this threshold, any applied fluctuating stress theoretically will never produce a fatigue failure. As the cyclic stress is increased, the time to failure is dramatically reduced because of the log function of time to fracture.

This curve is considered idealized because it is achieved only in vacuum. Once structures are placed in a service environment — even under atmospheric conditions — the S-N curve is described more often by curves like Figure 1–19, in which no true threshold exists and the application of a cyclic stress will ultimately induce failure. This AWS-X-modified curve is used in API RP2A for designing welded joints that are necessary for offshore platforms.

An example of fatigue life can be demonstrated using a frequency of 10 cpm, which produces the resulting scale of time at the top. Thus, at a stress range of 2,000 psi, a component would have a service life of 100 years, while doubling the stress range to 4,000 psi would reduce the life to approximately 50 years.

The fracture surface resulting from fatigue is quite distinctive and easily distinguishable. An example of a fatigue fracture is shown in Figure 1–20. The large semi-circular rings are referred to as beach marks and are characteristic of fatigue crack growth.

The threshold stress or fatigue limit for steels is approximately one-half of their tensile strength. This rule of thumb is useful for easy comparison of steels when no other data exists but should never be used for fatigue design.

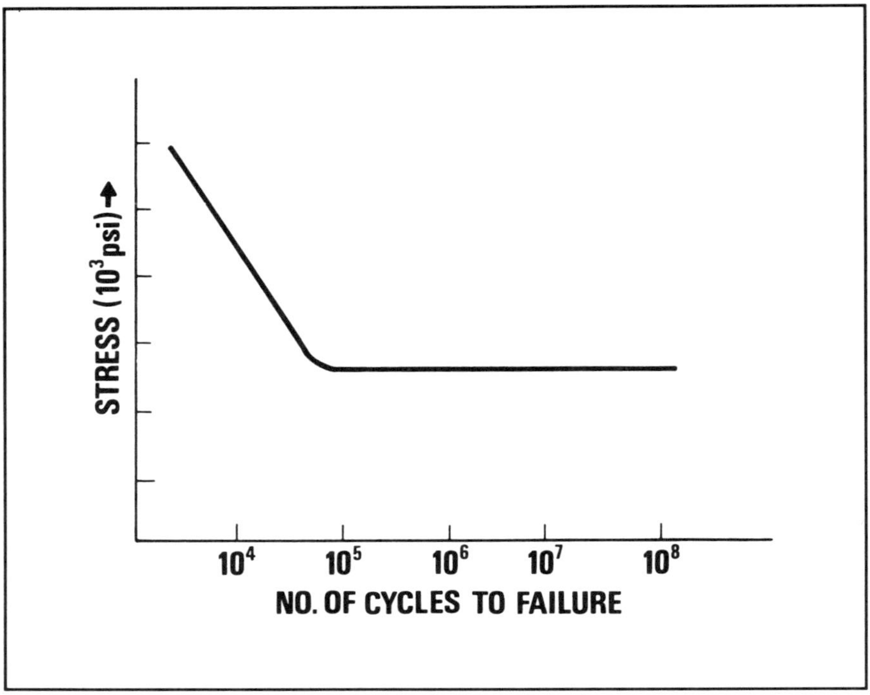

Figure 1–18 Typical S-N curve for fatigue of steel.

Fatigue can be caused by thermal fluctuations, ocean waves, wind, mechanical vibrations such as caused by operating machinery, and numerous other influences. This failure mechanism is quite common in the oilfield but is often misinterpreted. As in brittle fracture, the fracture surface may show bright fracture facets where the crack has followed the crystal boundaries or across the grains. This cracking has for many years been erroneously referred to as *crystallization*. This term is incorrect in this context since the metal is always crystalline below its melting point. The crystals are just more readily apparent because of the nature of the fracture and not the sudden formation of a crystalline structure.

Carbon and Alloy Steels

Steel is the most common alloy in the oilfield and as such deserves more attention to its specific behavior than other alloys. Steel is an alloy of carbon in iron with iron as the principal element. Carbon forms an interstitial solid solu-

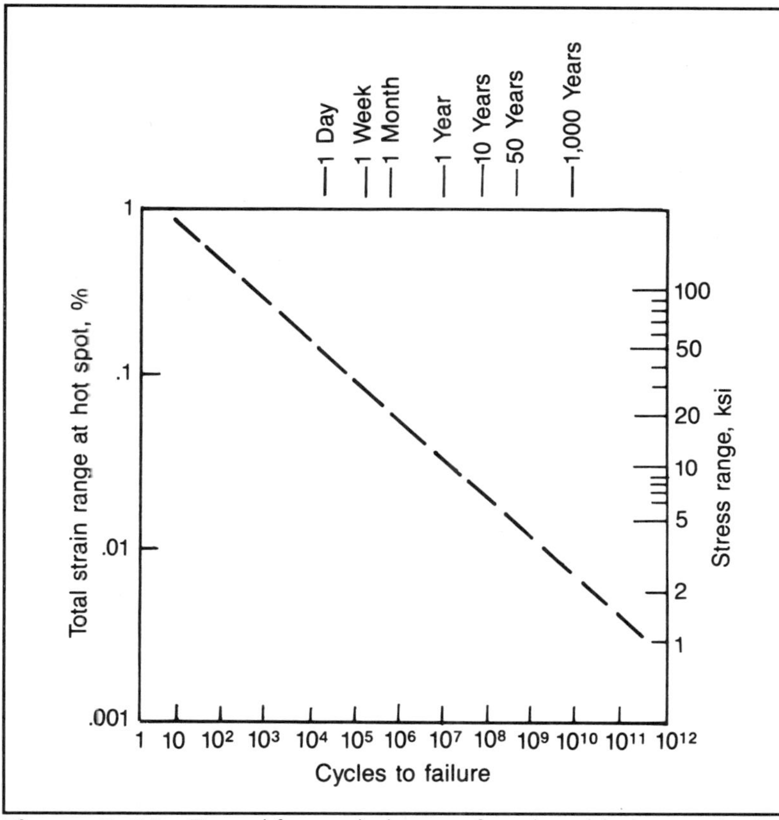

Figure 1–19 Typical fatigue behavior of steel, welded or notched and immersed in seawater (After API *RP 2A*).

tion in iron. Alloys with greater than 2.5% carbon are referred to as *cast irons* and will be dealt with in a later section.

Pure iron when cooled slowly from the liquid state goes through three changes in unit cell configuration. These allotropic transformations are liquid to bcc, bcc to fcc, and fcc to bcc in order of decreasing temperature. Thus, at room temperature the structure of pure iron is bcc. When carbon is added to iron in small quantities, there is no change in these transformations when the steel is slowly cooled. At this point, the discussion becomes more complex and is aided by the use of a phase diagram (see Fig. 1–21).

The *phase diagram* is a plot of the *equilibrium phases* present at a given temperature and composition. While these diagrams are literally packed with information, only their salient points will be presented for clarifying further topics. The diagram in Figure 1–21 consists of several single-phase regions (e.g., liquid, γ) and two-phase regions (e.g., $\alpha + Fe_3C$, γ + liquid, etc.). The two-

Figure 1–20 Fatigue fracture surface showing fatigue beach marks. Arrows indicate fracture origins.

phase regions represent equilibrium between the two adjoining single phases. It is apparent from the figure that when cooling from the liquid state along the 1.0% carbon composition line a region exists over a specific range of several hundred degrees, in which the two phases, liquid and delta iron (δ), are in equilibrium. The amount of each phase present at any temperature can be determined from the *Lever law* (an explanation of which can be found in any basic metallurgy book).

What is important to understand here is that a variety of phases or combinations of phases can be generated under equilibrium or near-equilibrium conditions. Since the steels commonly used in the oilfield contain less than 0.8% carbon (called *hypoeutectoid*), the remainder of this section will be devoted to that portion of the phase diagram to the left of the 0.3% carbon concentration line.

To demonstrate one of the uses of this diagram, consider a weld made in ASTM A 106 pipe that is nominally 0.30% carbon with a similar electrode. Since this diagram represents equilibrium conditions, the cooling rate must be quite slow; therefore, assume the weld is wrapped with insulation immediately after completion. Although this example is not close to equilibrium, it approximates the necessary conditions.

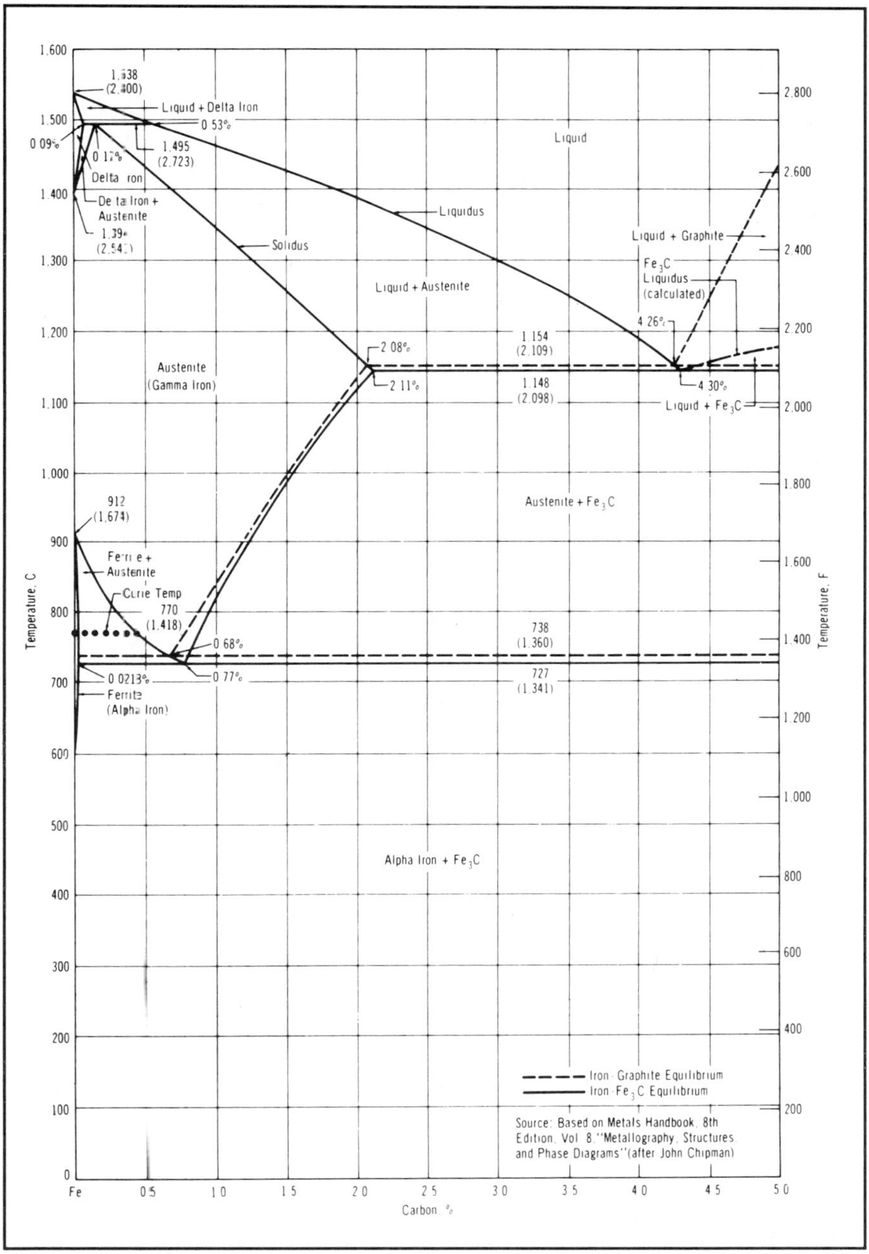

Figure 1-21 Iron-iron carbide equilibrium phase diagram (from *Metal Progress*).

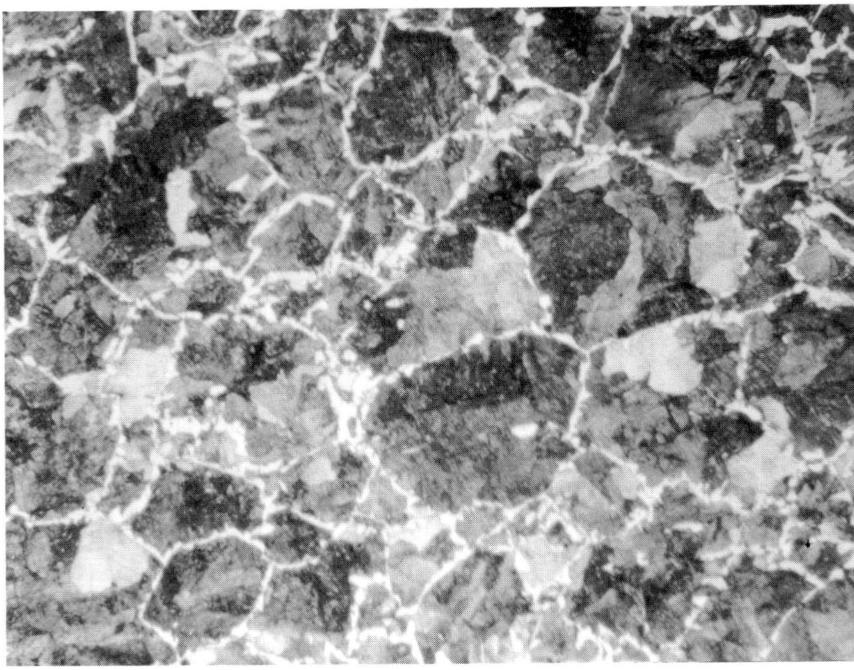

Figure 1–22 Example of pearlite microstructure from J-55tubing. Carbon content of this tubing was 0.52%. Etched in 2% nital and magnified 168 times.

As welding begins, the interface between the pipe wall and the welding rod becomes liquid. As cooling begins, the metal follows the 0.30% carbon line into a two-phase region of solid δ (delta iron) and liquid in which the first solid that forms is of the bcc structure because δ has a solubility for only up to 0.10% carbon and thus cannot absorb all of the carbon in the liquid and form a single phase of δ. Cooling below 2,723°F, the δ phase transforms to γ, or austenite as it is commonly called. Austenite and liquid now are in equilibrium until the temperature drops to approximately 2,690°F. At this point, all remaining liquid solidifies directly to austenite. Cooling through the austenite, which is now a solid single phase, produces no changes until a temperature of about 1,500°F is attained. At this time, the γ wants to transform to α (ferrite), but because of the excess carbon, since α has a solubility of only 0.02%, it must transform only partially, so γ and α are in equilibrium with each other. At 1,360°F, the remaining γ transforms to a two-phase constituent of ferrite plus cementite. The latter is of composition Fe_3C that contains 6.67 wt% carbon. This two-phase mixture, in which the cementite is present as alternating plates in a ferrite matrix, is termed pearlite. An example of pearlite from J-55 tubing, is shown in Figure

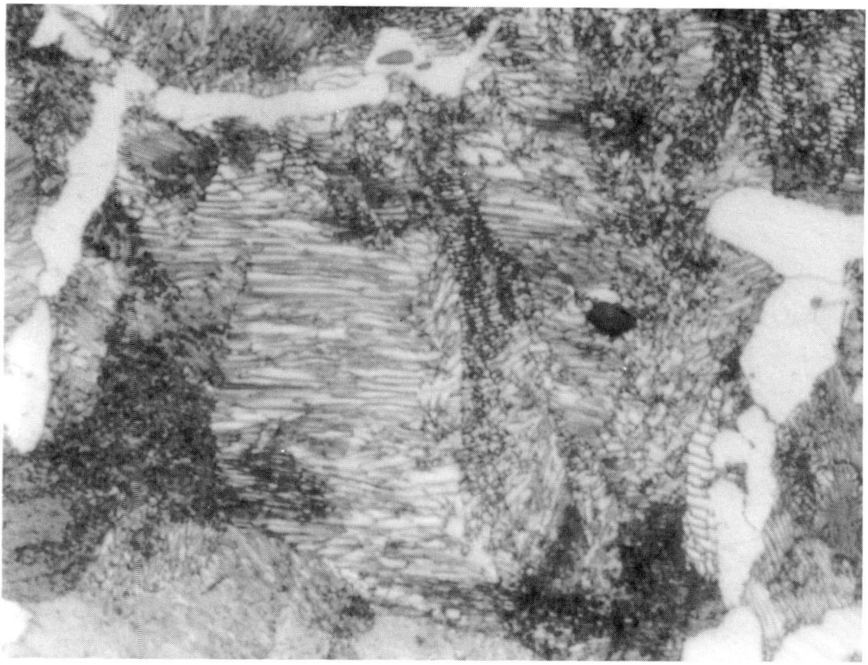

Figure 1–23 Same as Figure 1–22 but magnified 1,250 times to reveal the cementite (black lines) and ferrite (white areas).

1–22 and, at a higher magnification, in Figure 1–23, reveals the separate cementite plates in the ferrite matrix.

To this point, the phases formed under near-equilibrium conditions have been discussed, but if cooling is faster, what phases appear under nonequilibrium cooling? Consider the A 106 tube, but this time, instead of being wrapped with an insulating blanket, it is allowed to cool in air. No significant changes occur above the austenite region so only those below are considered. Again, the ferrite forms, but the faster cooling or quench produces a much finer pearlite in which the spacing of the cementite plates is closer. If the cooling is increased, another mixture of ferrite and carbide is produced, but with an entirely different appearance, and is called bainite. Figure 1–24 is an example of bainite. Increasing the cooling rate by using a water or oil quench produces martensite, a single-phase, highly strained matrix (see Fig. 1–25).

These various microstructures, which produce a variety of properties, are the basis for the heat treatment of steels. However, it is evident from Figure 1–21 that most of them cannot be predicted from the equilibrium phase diagram. Therefore, a different method must be adopted to describe and predict these

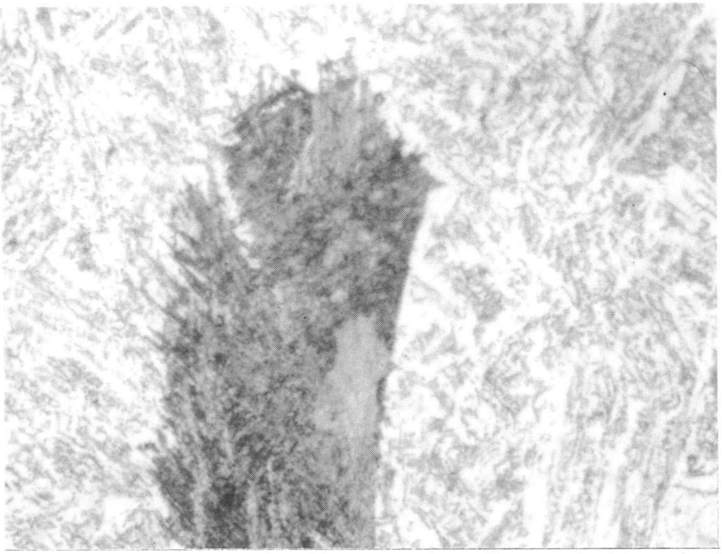

Figure 1–24 Example of bainite in low-alloy steel. Magnified 1000 times.

microstructures. This method will be dealt with in the next section on heat treatment.

Thus far, only the iron-carbon system has been discussed, but steels typically contain many other elements added on purpose or inherited from the steel-making process. Oxygen, sulfur, phosphorous, nitrogen, tin, antimony, and arsenic are the principal impurities. These elements and others are responsible for the diminished resistance of steels to impact loading, high-temperature service, fatigue, etc. To overcome these detrimental aspects or enhance certain properties, alloying elements are added.

Silicon and aluminum are excellent deoxidizers, tying up oxygen as inclusions that may further improve the mechanical properties of a steel since these inclusions tend to pin grain boundaries and inhibit grain growth. Silicon also increases strength in quenched and tempered steels and moderately improves hardenability, which will be discussed later in this chapter. Titanium is another deoxidizer but is primarily added because of its ability to strengthen the matrix by precipitating titanium carbides and nitrides.

Molybdenum significantly improves hardenability and high-temperature strength and increases steel's resistance to softening when tempered. When added to austenitic stainless steels, it increases resistance to pitting attack. Nickel is beneficial for improving toughness of a steel and hardenability. When Ni is added to high-chromium steels it stabilizes the austenite phase below room

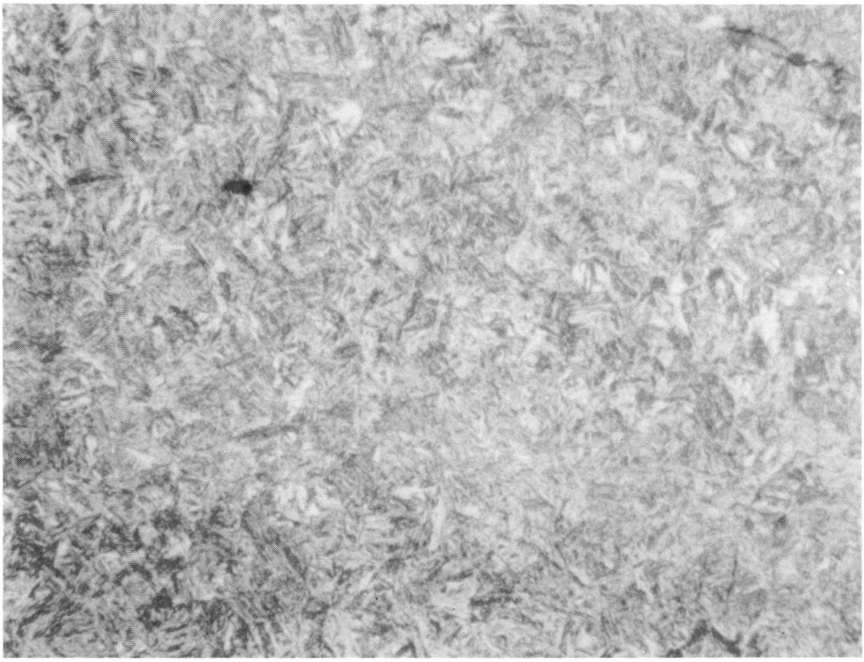

Figure 1–25 AISI 4130 low-alloy steel water-quenched from 1,600°F. Etched to show the fine martensite laths.

temperature. Manganese also promotes austenite stabilization but is primarily added for increasing hardenability at low cost and tying up sulfur to prevent free sulfur or iron sulfides from being present since they are detrimental to steel.

Chromium is added to improve corrosion and oxidation resistance and for its hardenability contribution. At higher concentrations, it improves high-temperature strength and abrasion resistance.

Tungsten provides increased strength as a carbide former and resistance to temper softening much as molybdenum does. It also contributes strongly to hardenability and improves the high-temperature strength of steels.

Columbium and vanadium are strong carbide formers and, because of their fine dispersion as carbides help pin grain boundaries thereby producing a fine grain size. They also contribute significantly to hardenability and resistance to softening during tempering.

The addition of alloying elements (up to 5 wt%) categorizes the steel as low alloy, whereas above this value, the steel is called an alloy steel.

With such a variety of elements and different possible combinations, organized numerical codes for often-used alloys were devised. While there are numerous alloy designation systems, only a few are commonly used in the petroleum industry. These are the designation systems of the American Society of Testing and Materials (ASTM), the American Iron and Steel Institute (AISI), and the Society of Automotive Engineers (SAE). The last two, AISI and SAE, are almost identical and will be referred to as AISI for ease of discussion. In addition, although it does not have an alloy system, the American Petroleum Institute (API) has a series of specifications covering products of nominal chemistry that are widely used in the industry. The American Society of Mechanical Engineers (ASME) also has a much-used system of alloy designations, but since it is almost identical to the ASTM, it will not be discussed further.

To explain the systems briefly, the ASTM designation for A 106 Grade B pipe is properly A-106-80 Grade B. The A prefix indicates a ferrous alloy; a B in the prefix is for nonferrous alloys. The 106 has no particular significance, while the 80 is the year of last revision of the standard. The AISI system has an advantage over the ASTM designation in that the composition is indicated in the designation itself. There is a four-number designation for AISI alloys. The first two numbers describe the major alloying elements and the last two, the nominal weight percent of carbon in 0.01% increments. The series is as follows:

10XX	Plain carbon steel
11XX	Free cutting steel: high sulfur
12XX	Free cutting steel: high sulfur and/or phosphorus
13XX	Manganese steel: Mn 1.75%
15XX	Plain carbon steel: Mn 0.75-1.65%
23XX, 25XX	Nickel steels: Ni 3.5-5.0%
31XX, 32XX, 33XX	Nickel-chromium steels: Ni 1.25-3.5%, Cr 0.65-1.57%
40XX	Molybdenum steels: Mo 0.20-0.30%
41XX	Chromium-molybdenum steels: Cr 0.8-1.1%, Mo 0.15-0.25%
43XX	Ni 1.65-2.00, Cr 0.40-0.90, Mo 0.20-0.30
47XX	Ni 0.90-1.20, Cr 0.35-0.55, Mo 0.15-0.25
81XX	Ni 0.20-0.40, Cr 0.35-0.55, Mo 0.08-0.15
86XX	Ni 0.40-0.70, Cr 0.40-0.60, Mo 0.15-0.25
87XX	Ni 0.40-0.70, Cr 0.40-0.60, Mo 0.20-0.30
88XX	Ni 0.40-0.70, Cr 0.40-0.60, Mo 0.30-0.40
94XX	Ni 0.30-0.60, Cr 0.30-0.50, Mo 0.08-0.15
98XX	Ni 0.85-1.15, Cr 0.70-0.90, Mo 0.20-0.30

46XX, 48XX	Nickel-molybdenum steels: Ni 1.55-3.5%, Mo 0.20-0.30%
50XX, 51XX	Chromium steels: Cr 0.25-1.05%
61XX	Chromium-vanadium steels: Cr 0.8-0.95%, V 0.1-0.15%
72XX	Tungsten-chromium steels: W 1.75%, Cr 0.75%
92XX	Silicon-manganese steels: Si 1.2-2.2%, Mn 0.65-0.87%
XXBXX	Boron steels
XXLXX	Leaded steels

Thus, an often-used oilfield steel such as AISI 4140 would contain nominally 1% chromium, 0.2% molybdenum, and 0.4% carbon. Because the use of AISI numbers allows easy recognition of the basic chemistry of the alloy, people commonly try to correlate API products with an AISI designation. In the lower grades of tubing, casing, drillpipe, and line pipe, this is almost impossible because of the lack of specific chemistry requirements for these API grades. For the special grades, such as restricted-yield tubing and casing and higher-strength grades, the chemistry gap is somewhat reduced but never to the point that an actual AISI designation can be assigned. This is one of the major problems of oilfield metallurgy. For example, the chemistry of J-55 tubing can vary significantly from mill to mill and still meet the required mechanical properties. These variations will become more evident in Chapter Four when nominal compositions are discussed.

Recently, the ASTM has established the Unified Numbering System (UNS) to consolidate the various alloy systems into one for ease of application. Table 1–2 shows the organization of the UNS designation system.

TABLE 1–2

ASTM Unified Numbering System

AXXXXX	Aluminum and Aluminum Alloys
CXXXXX	Copper and Copper Alloys
EXXXXX	Rare Earth and Similar Metals and Alloys
FXXXXX	Cast Irons
GXXXXX	AISI and SAE Carbon and Alloy Steels
HXXXXX	AISI and SAE H-Steels
JXXXXX	Cast Steels (Except Tool Steels)
KXXXXX	Miscellaneous Steels and Ferrous Alloys

LXXXXX	Low Melting Metals and Alloys
MXXXXX	Miscellaneous Non-Ferrous Metals and Alloys
NXXXXX	Nickel and Nickel Alloys
PXXXXX	Precious Metals and Alloys
RXXXXX	Reactive and Refractory Metals and Alloys
SXXXXX	Heat and Corrosion Resistant Steels (Including Stainless), Valve Steels, and Iron-Base "Superalloys"
TXXXXX	Tool Steels, Wrought and Cast
WXXXXX	Welding Filler Metals
ZXXXXX	Zinc and Zinc Alloys

Heat Treatment of Steel

The heat treatment of steel most often involves heating into the austenite region shown in Figure 1–21 and, depending on the desired properties, cooling at a predetermined rate to room temperature. The faster the quench or cooling rate, the finer the microstructure and the higher the strength. Also, with increasing strength, the ductility decreases as dramatically. To regain some of the ductility, a carbon steel is reheated to temperatures below 1,360°F, which is called the lower critical temperature or Ac_1. This form of heat treatment is referred to as tempering. The Ac_1 is not a constant temperature but is different for every alloy system and is highly dependent on the rate of heating. Diagrams that show the transformation characteristics of steel and the phases formed at time-temperature-transformation are called (TTT) diagrams. Figure 1–26 is an example of one of these diagrams. These diagrams are useful from a practical standpoint since they represent those phases or mixtures formed under nonequilibrium conditions. They also describe the kinetics of transformations, which, of course, an equilibrium phase diagram cannot. The basis of the TTT diagram is transformation under isothermal conditions. Thus, steel must be cooled from the critical temperature instantaneously to a temperature of interest and then held constant at that temperature. The various lines represent the start and finish of the transformation to pearlite, ferrite, bainite, or martensite.

To illustrate the concept and use of the diagram, assume steel is instantly cooled to 1,200°F from the austenite region and held at this temperature for 1 min. Lines that are crossed when the temperature is held constant are the austenite plus ferrite, the pearlite start, the dotted line indicating 50% completion of the pearlite transformation, and the last line, the pearlite end. Thus, after 1 min the austenite has completely transformed to pearlite and ferrite. The microstructure will look similar to that shown in Figure 1–22 with ferrite at the prior austenite grain boundaries and pearlite within the grains. Although there are many reactions and combinations that can be described by these diagrams, one of their shortcomings is the instantaneous or idealized quench to the requisite temperature. This is especially inconsistent with reality when trying to

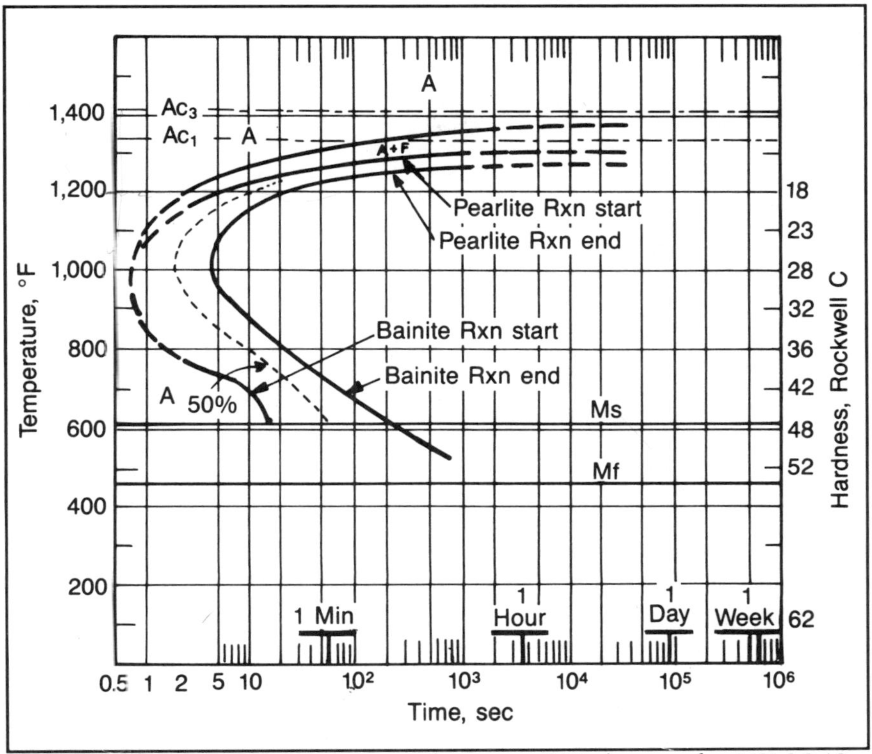

Figure 1–26 Time-temperature-transformation (TTT) diagram for AISI 1040 steel. Approximate hardness values are shown at the right for isothermal treatment at the respective temperature.

achieve the lower-temperature transformation products such as martensite.

A more useful diagram to predict the results of heat treatment that compensates for this quench-time inconsistency is a CCT (Continuous Cooling Transformation) *diagram*. Figure 1–27 shows such a diagram for a medium-carbon steel. There are three regions: ferrite + pearlite, bainite, and martensite. The ferrite + pearlite region is the envelope to the right with the bainite envelope in the middle. The left open sided envelope is for martensite. The one and ninety-nine represent 1% and 99% transformation to the specific product. The values in the small circles at the bottom of the diagram provide hardness information from the resulting microstructure. It is evident by the DPH (diamond pyramid hardness) 620 value that rapid cooling from the austenite at 1,360°F produces almost 100% martensite, whereas very slow cooling produces a structure composed of ferrite and pearlite. The cooling line representing a hardness of DPH 351 shows a partial transformation to bainite while the remaining austenite would transform to martensite.

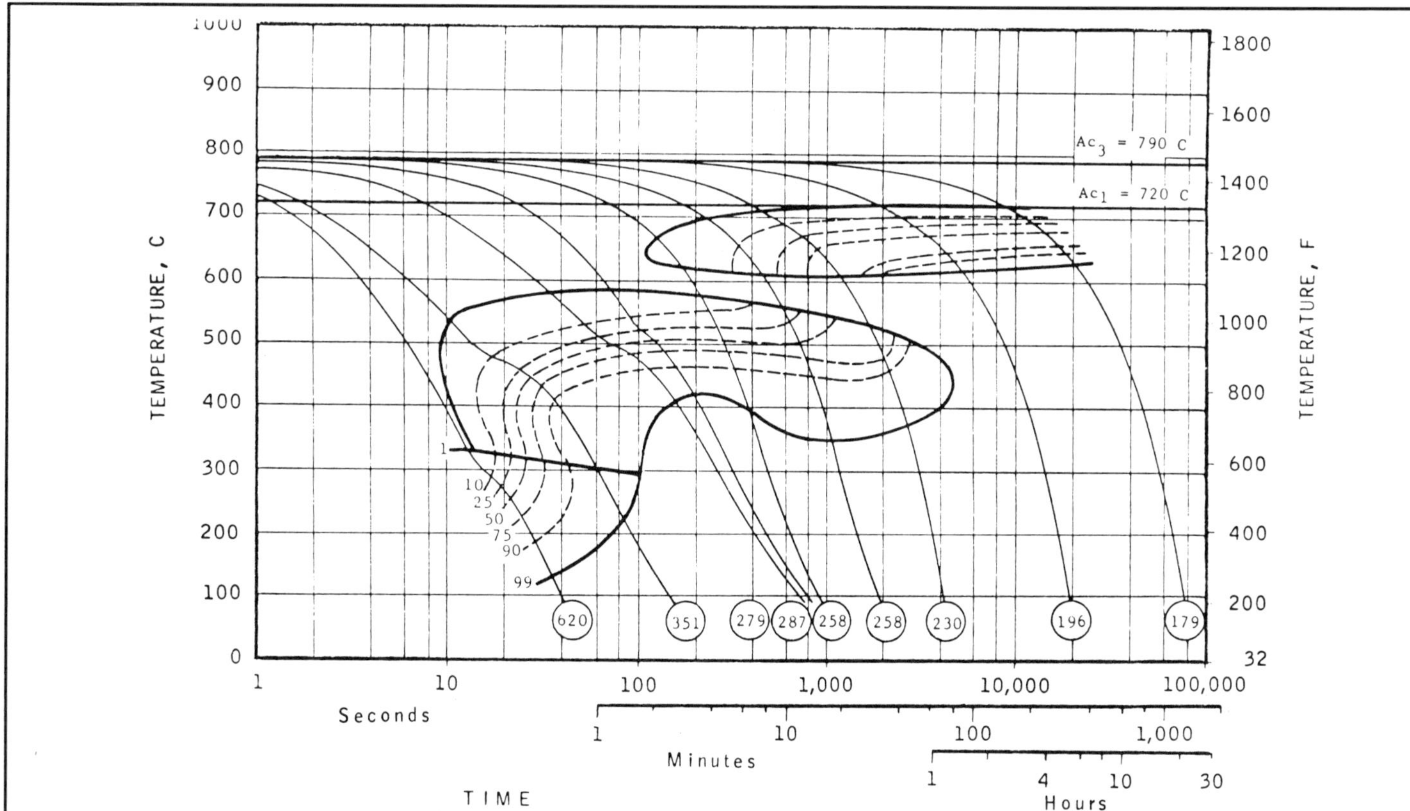

Figure 1–27 Continuous cooling transformation (CCT) diagram for AISI 1040 steel modified with 0.53% molybdenum (After Cias).

At this point, it is advantageous to define certain heat-treating terminology. All of these definitions assume the steel is already heated into the austenite region. If cooling occurs in the furnace very slowly, the treatment is referred to as *annealing* and produces a soft steel with carbides that are spherical in shape or a very coarse pearlite. *Normalizing* requires removing the steel from the furnace and allowing it to cool in air — usually still air. Quenching is usually achieved with anything faster than still air such as forced air, water, brine, or oil and is most often accomplished by immersion.

The major point of both the TTT and CCT diagrams to be gained from this discussion is that transformations are a function of both cooling rate and chemistry. This is important in the manufacture of heavy-wall castings or forgings for valve bodies or heavy-wall pipe for deep wells. To achieve high strength, the alloy must first form martensite. If the necessary cooling rate by quenching cannot be achieved in these heavy-walled products, then alloying elements must be added that will permit slower cooling and still produce a martensitic structure.

The ability to produce martensite either by rapid cooling or by alloying is referred to as *hardenability*. This term should not be confused with hardness, which is a measure of resistance to deformation. The two are related in that the hardenability of a steel determines the depth and distribution of hardness resulting from quenching from the austenite region. Hardenability is a function of carbon content, alloying, austenite grain size, time and temperature during austenizing, and cooling rate. Elements that increase hardenability are manganese, molybdenum, chromium, silicon, nickel, and boron to name the major ones.

In many applications, such as quenched and tempered oilfield tubulars, the ability to produce uniform hardness consistently across the wall thickness requires close controls on the steel chemistry used for manufacturing. This is accomplished by ordering H-steels (e.g., 4140H), which are required to have hardness limits at certain depths that are stricter than for normal carbon and low-alloy steels. This is shown schematically in Figure 1–28. The H-band has a tighter spread on hardness and thus more uniform response to heat treatment and more uniform properties than a standard steel. The 1340H steel is sometimes used for tool joints in heavy-wall drillpipe, and the 4140H is used in numerous oilfield applications. Notice that both steels display a declining hardness with depth into the steel. This limit to the depth of hardening has serious ramifications on heavy wall equipment and the ability to contain high pressures.

Once the martensitic structure is obtained, it is frequently tempered to regain some ductility although there is also an accompanying loss of strength. Figures 1–29 and 1–30 show the loss in hardness and yield strength with temperating temperatures for AISI 4130 steel and its modifications. As can be seen by the elongation data (Fig. 1–30), which are a good measure of ductility, from a 300°C temper to a 700°C temper the elongation doubles while the yield strength falls by more than half. The microstructure also changes significantly and is shown in Figure 1–31. Figure 1–31a shows a lightly tempered martensite with a high degree of directionality. The needles, or laths as they are called (Fig. 1–31a) are remnants of the quenched martensite. Figure 1–31b, however, shows little relation to the original martensite and is more uniform and nondirectional in

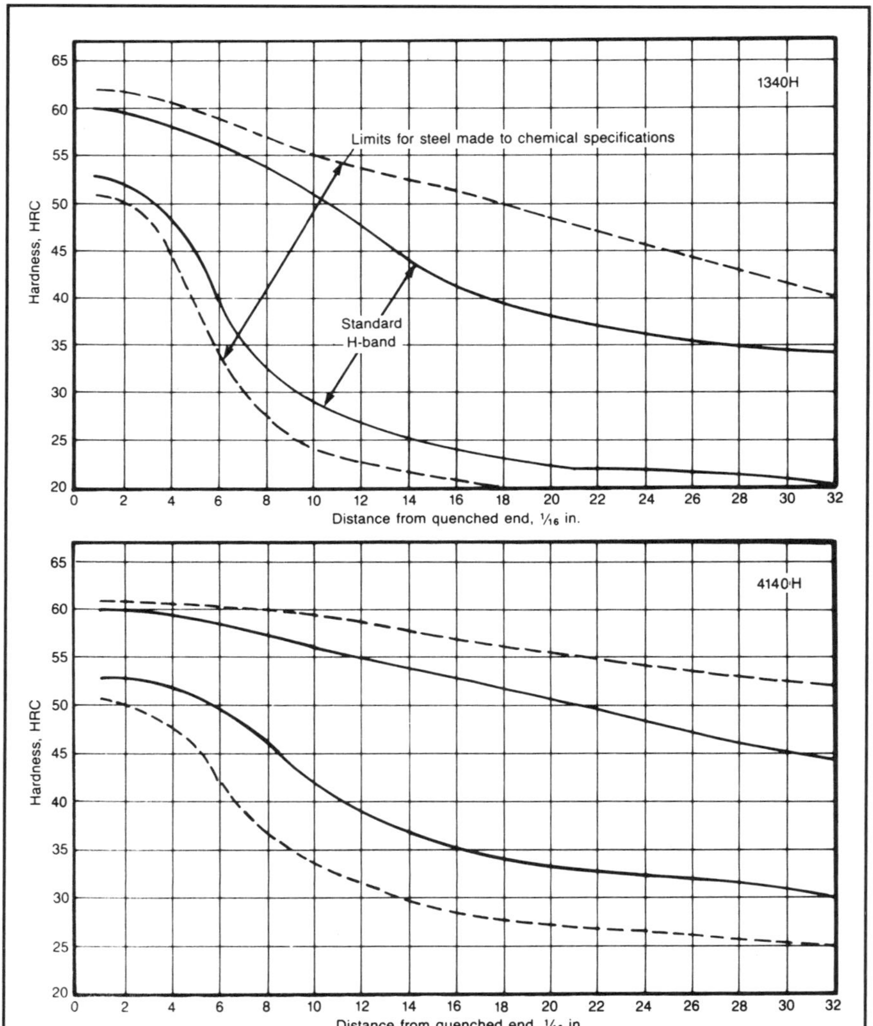

Figure 1–28 Comparison of standard H–bands and wider limits for similar steels made to AISI specifications (After *Jatczak, Metals Handbook,* 1978).

appearance. These differences in microstructure are the major factors underlying the strength and ductility of a steel. It is this relationship between structure and properties of a metal that forms the basis of metallurgy and should be the important concept gained from this chapter.

Cast Irons

Cast irons are alloys of iron and carbon (greater than 2.0%) with silicon. Depending on the properties desired, other elements such as manganese, phosphorus, nickel, cerium, and magnesium are added. The main classes of cast iron

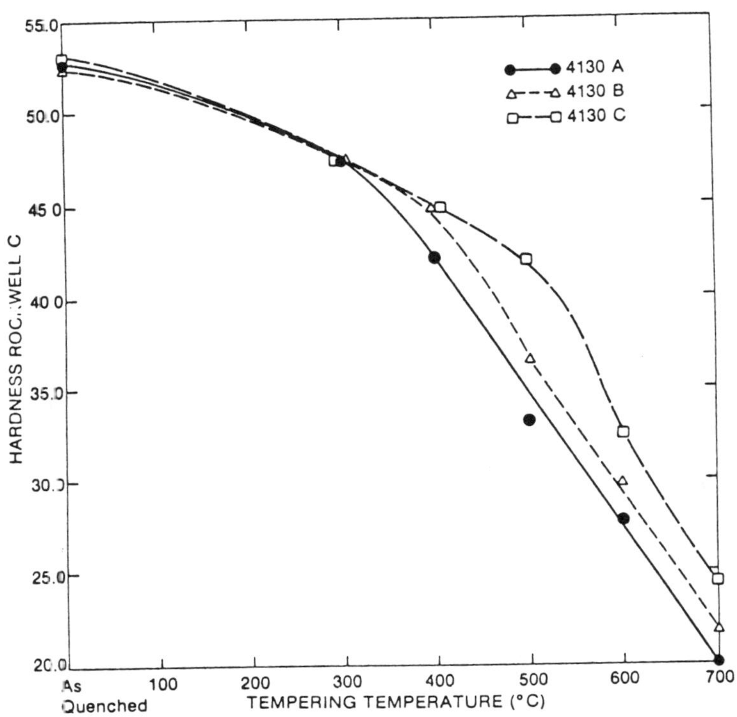

Figure 1–29 Reduction in hardness with increasing tempering temperature for AISI 4130 steel and two modifications with 0.50% molybdenum (4130B) and 0.75% molybdenum (4130C).

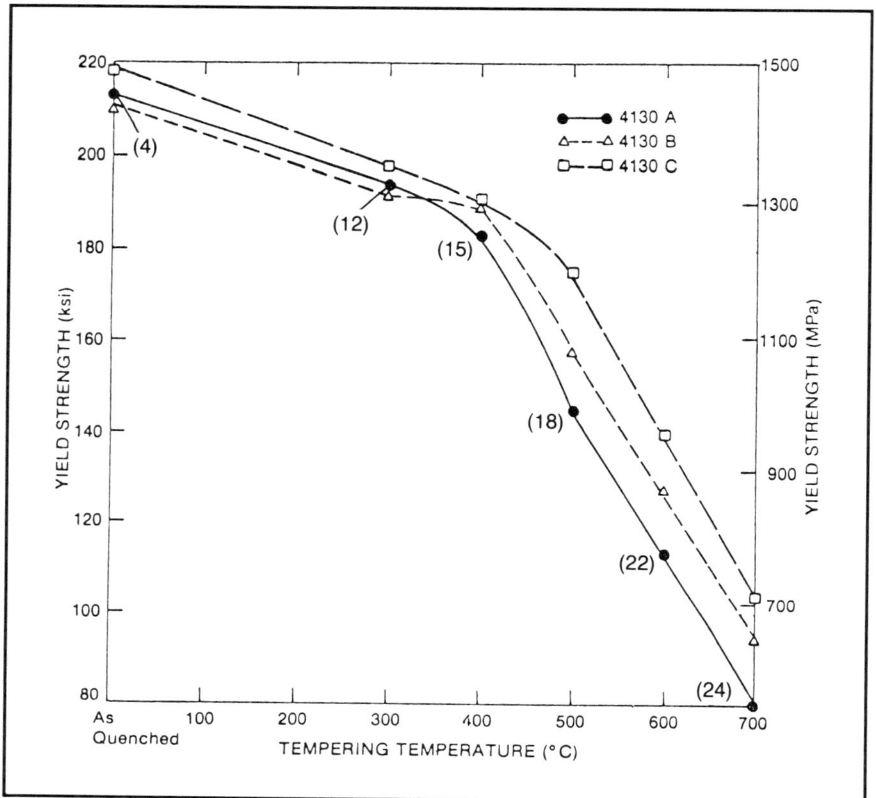

Figure 1–30 Reduction in yield strength with increasing tempering temperature for the same series of alloys shown in Figure 1–29. Values in parenthesis are % elongation in 2 in.

are gray, white, malleable, and ductile. As in steels, the variations in structure of these alloys produce a variety of properties.

Gray iron has flakes of graphite in a ferrite matrix (Fig. 1–32), which gives this alloy its characteristic gray appearance when fractured. Ferritic gray cast iron has low strength but is less costly compared to other cast irons. Gray cast irons with carbon contents less than 4.3% are termed *hypoeutectic* and those greater than 4.3%, *hypereutectic*. The former are noted for their high strength, while the latter display excellent thermal shock resistance. The presence of graphite in flake form enhances the machinability of these alloys. However, it also significantly reduces the ductility of the gray cast irons because of the strong notch effect it presents.

Gray irons are designated by the ASTM with a two-digit system that corresponds to the minimum tensile strength in 1,000-psi increments. Thus, a class

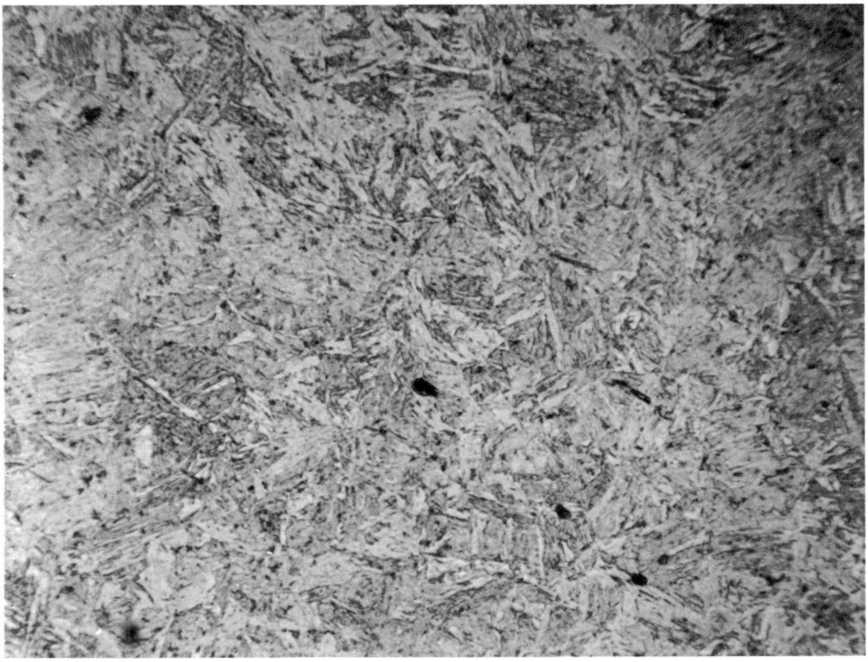

Figure 1–31a Tempered martensite of 4130 steel, tempered 1 hr at 300°C. Nital etch, magnified 690 times.

No. 40 cast iron would have a minimum tensile strength of 40,000 psi.

White cast iron is distinguished from the other cast irons by its lack of graphite (see Fig. 1–33). Because of the high concentration of iron carbides, this cast iron is quite hard (BHN 400–450) and brittle, which makes it a good alloy for service in highly abrasive environments. White irons alloyed with nickel and chromium and quenched to martensite can attain hardnesses of 600–725 Brinell.

Malleable cast irons are made from white iron castings that are heated to 1,500–1,750°F, held for a specified length of time depending on thickness and composition, then slowly cooled through a temperature range of 1,400–1,300°F. There are three types of malleable iron: ferritic, pearlitic (martensitic), and alloy. The first has a ferritic microstructure with carbon present in nodule form; the second type has carbon retained both in nodules and in cementite or martensite; and the alloy malleable iron has carbon in the form of graphite nodules as in the ferritic type. Malleable cast irons are used where both high strength and wear resistance are needed.

Adding magnesium or cerium to molten gray iron causes the graphite flakes to spheroidize (see Fig. 1–34), promoting increased ductility compared to gray cast iron. Ductile irons have greater impact resistance than other cast irons and, like

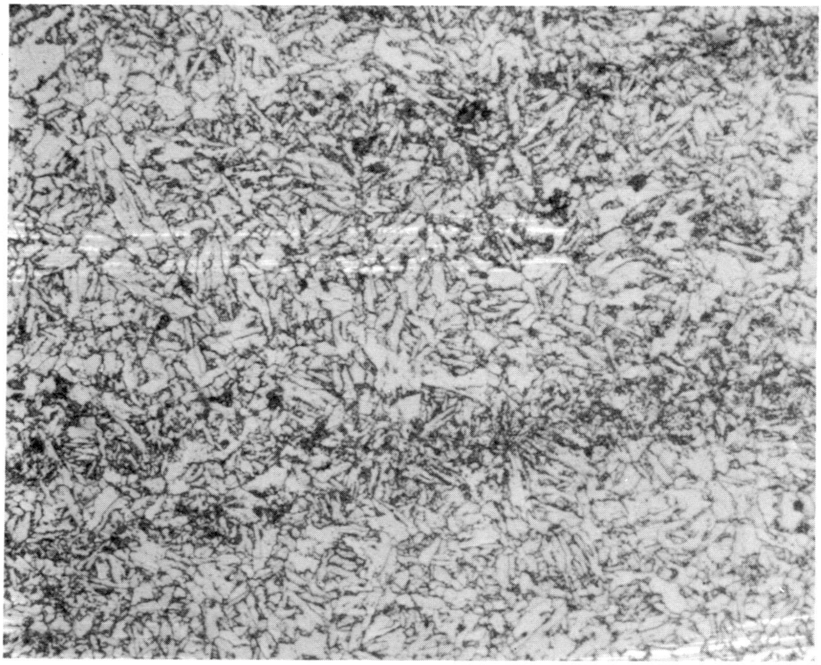

Figure–31b Highly tempered structure of 4130 steel. Tempered 1 hr at 700°C. Nital etch, magnified 368 times.

steel, can be heat-treated to tensile strengths on the order of 180,000 psi. An important property of ductile iron is that it behaves elastically, like steel, when stressed and thus follows Hooke's law. Gray iron, on the other hand, does not. Because of the sharp notch effect of the graphite, it shows only elastic behavior under light loads.

Ductile irons are designated by ASTM A 536 by a series of three numbers: for example, 60–40–18 in which 60 represents a minimum tensile strength of 60,000 psi, 40 a minimum yield strength of 40,000 psi, and 18 a minimum elongation of 18%. Ductile iron is used in applications in which high strength plus good ductility are required.

Tensile strengths of cast irons in the four major categories fall in the following ranges: 20–80 ksi for gray irons, 13–90 ksi for white irons, 50–100 ksi for malleable irons, and 55–175 ksi for ductile irons. Hardness values of cast irons are 140–350 Brinell for gray irons, up to 600 for white irons, 110–270 for malleable irons, and 130–300 for ductile irons. A simple relationship between hardness and tensile strength does not exist for cast irons as it does for steel because of the more heterogeneous microstructure prevalent in the cast irons.

In the oilfield, cast irons are used in valves, pumps, packers, bridge plugs,

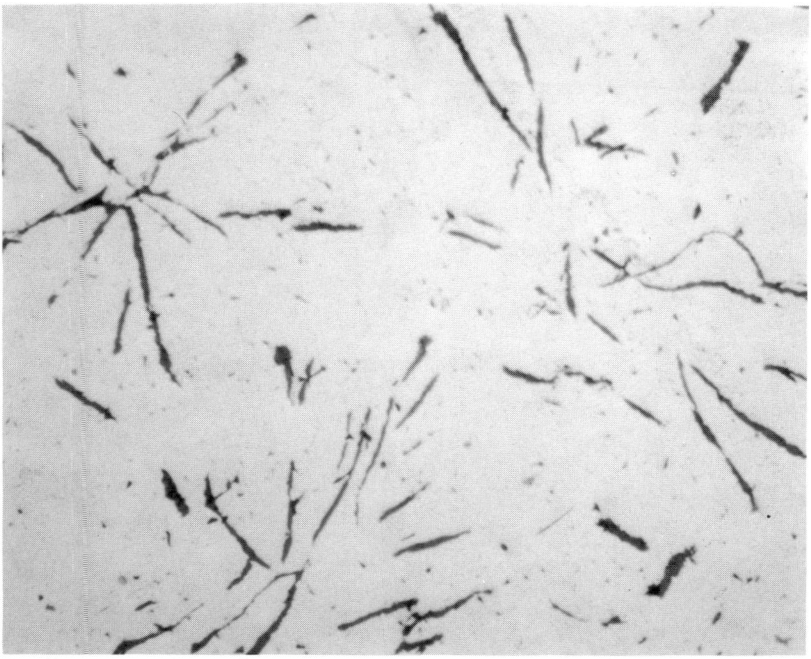

Figure 1–32 Gray cast iron showing graphite flakes. No etch, magnified 175 times.

and other assemblies. It should be appreciated that selection of cast irons requires a knowledge of the intended application since, in many ways, cast irons are less forgiving than steels.

Stainless Steels

Until the last 5 years stainless steels were not widely used in petroleum production except in very specialized applications such as trim on valves or parts in downhole tools. Recently, drilling and production in aggressive environments have led to the need for highly alloyed materials such as stainless steels, nickel-base alloys, cobalt-base alloys, and titanium alloys.

Chromium is the element that gives stainless steel the stainless quality. To be classed as stainless, steels must contain 12% chromium or more. Nickel added in excess of 6% increases corrosion resistance and substantially improves mechanical and fabricating properties. Molybdenum increases pitting resistance and enhances high-temperature strength.

Stainless = 12+% Chromium

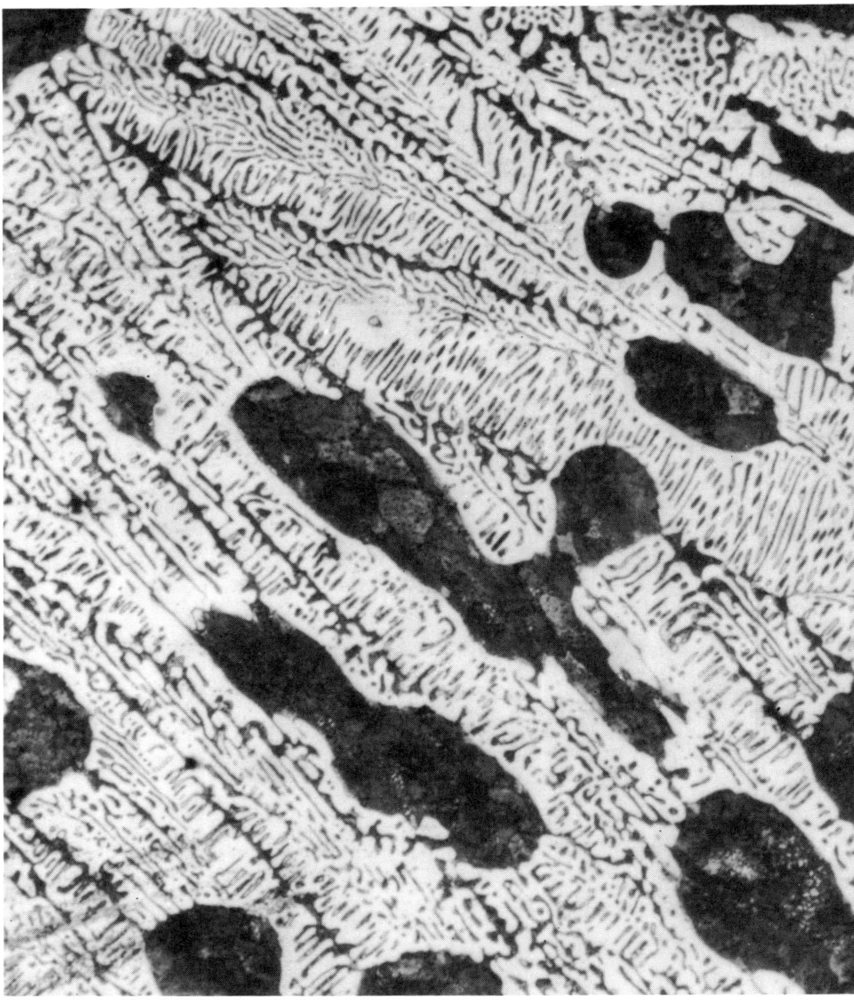

Figure 1–33 Hypoeutectic white cast iron, large dark areas of pearlite and speckled areas of ledeburite. Nital etch, magnified 52 times.

Stainless steels are classified by the AISI according to metallurgical structure as shown in Table 1–3. The 200 and 300 series are austenitic while the 400 series is either martensitic or ferritic. Another series is the precipitation-hardening stainless steels. These steels have a different designation such as 17–4 PH, indicating that they contain nominally 17% Cr–4% Ni. Lastly, there are a group of alloys referred to as duplex stainless steels that combine the austenite and ferrite phases in one alloy.

The 200 and 300 series are not hardenable by heat treatment, but only by

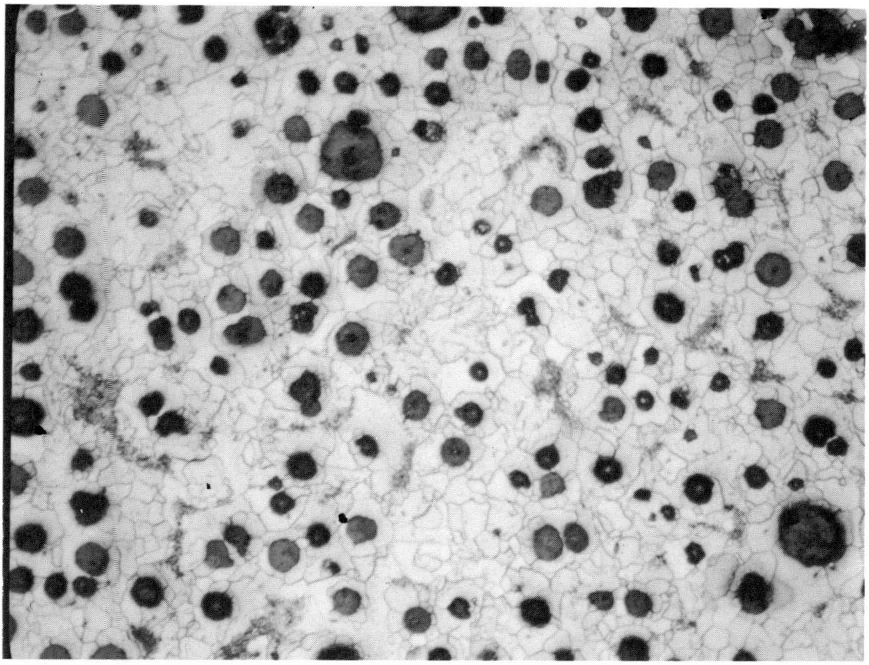

Figure 1–34 Ductile iron showing graphite nodules in a ferrite matrix. Nital etch, magnified 145 times.

cold work. The mechanical properties of the austenitics are classified according to minimum tensile and yield strength into four categories called tempers.

Tempers	*Min. Tensile Strength, psi*	*Min. Yield Strength, psi*
1/4 hard	125,000	75,000
1/2 hard	150,000	110,000
3/4 hard	175,000	135,000
Fully hard	185,000	140,000

The austenitics in general have superior corrosion resistance compared to the ferritic and martensitic grades.

The ferritic stainless steels cannot be hardened by heat treatment and are used primarily in the annealed condition or with a minor amount of cold work. The martensitic alloys, on the other hand, have a high degree of hardenability and are widely used where both strength and corrosion resistance are needed. The precipitation-hardened stainless steels achieve their strength through heat treatment by two means. The first is transformation to martensite while the

Table 1–3 Compositions of AISI standard grades of stainless steels

UNS designation	AISI type	Composition, %(a) C	Mn	P	S	Si	Cr	Ni	Mo	Others
Austenitic grades										
S20100	201	0.15	5.60–7.50	0.06	0.03	1.00	16.00–18.00	3.50–5.50	...	0.25N
S20200	202	0.15	7.50–10.0	0.06	0.03	1.00	17.00–19.00	4.00–6.00	...	0.25N
S20500	205	0.12–0.25	14.00–15.50	0.03	0.03	0.50	16.50–18.00	1.00–1.75	...	0.32–0.40N
S30100	301	0.15	2.00	0.045	0.03	1.00	16.00–18.00	6.00–8.00	...	...
S30200	302	0.15	2.00	0.045	0.03	1.00	17.00–19.00	8.00–10.00	...	...
	302B	0.15	2.00	0.045	0.03	2.00–3.00	17.00–19.00	8.00–10.00	...	...
S30300	303	0.15	2.00	0.2	0.15	1.00	17.00–19.00	8.00–10.00	0.60	...
	303Se	0.15	2.00	0.2	0.06	1.00	17.00–19.00	8.00–10.00	...	0.15Se min
S30400	304	0.08	2.00	0.045	0.03	1.00	18.00–20.00	8.00–10.50	...	...
S30403	304L	0.03	2.00	0.045	0.03	1.00	18.00–20.00	8.00–12.00	...	...
S30430	304HQ	0.08	2.00	0.045	0.03	1.00	17.00–19.00	8.00–10.00	...	3.00–4.00C
	304N	0.08	2.00	0.045	0.03	1.00	18.00–20.00	8.00–10.50	...	0.10–0.16N
S30500	305	0.12	2.00	0.045	0.03	1.00	17.00–19.00	10.50–13.00	...	...
S30800	308	0.08	2.00	0.045	0.03	1.00	19.00–21.00	10.00–12.00	...	...
S30900	309	0.2	2.00	0.045	0.03	1.00	22.00–24.00	12.00–15.00	...	...
	309S	0.08	2.00	0.045	0.03	1.00	22.00–24.00	12.00–15.00	...	...
S31000	310	0.25	2.00	0.045	0.03	1.50	24.00–26.00	19.00–22.00	...	...
	310S	0.08	2.00	0.045	0.03	1.50	24.00–26.00	19.00–22.00	...	...
S31400	314	0.25	2.00	0.045	0.03	1.50–3.00	23.00–26.00	19.00–22.00	...	...
S31600	316	0.08	2.00	0.045	0.03	1.00	16.00–18.00	10.00–14.00	2.00–3.00	...
	316F	0.08	2.00	0.2	0.10 min	1.00	16.00–18.00	10.00–14.00	1.75–2.50	...
S31603	316L	0.03	2.00	0.045	0.03	1.00	16.00–18.00	10.00–14.00	2.00–3.00	...

	316N	0.08	2.00	0.045	0.03	1.00	16.00–18.00	10.00–14.00	2.00–3.00	0.10–0.16N
S31700	317	0.08	2.00	0.045	0.03	1.00	18.00–20.00	11.00–15.00	3.00–4.00	...
S31703	317L	0.03	2.00	0.045	0.03	1.00	18.00–20.00	11.00–15.00	3.00–4.00	...
S32100	321	0.08	2.00	0.045	0.03	1.00	17.00–19.00	9.00–12.00	...	Ti.5xC m.
	329	0.10	2.00	0.04	0.03	1.00	25.00–30.00	3.00–6.00	1.00–2.00	...
	330	0.08	2.00	0.04	0.03	0.75–1.50	17.00–20.00	34.00–37.00	...	0.10Ta.0.
S34700	347	0.08	2.00	0.045	0.03	1.00	17.00–19.00	9.00–13.00	...	Nb:10xC
S34800	348	0.08	2.00	0.045	0.03	1.00	17.00–19.00	9.00–13.00	...	Nb:10xC
	384	0.08	2.00	0.045	0.03	1.00	15.00–17.00	17.00–19.00	...	...
Ferritic grades										
S40500	405	0.08	1.00	0.04	0.03	1.00	11.50–14.50	...	...	0.10–0.30
S40900	409	0.08	1.00	0.045	0.045	1.00	10.50–11.75	...	...	Ti:6xC-0
S42900	429	0.12	1.00	0.04	0.03	1.00	14.00–16.00	...	...	...
S43000	430	0.12	1.00	0.04	0.03	1.00	16.00–18.00	...	...	...
	430F	0.12	1.25	0.06	0.15	1.00	16.00–18.00	...	0.60	...
	430FSe	0.12	1.25	0.06	0.06	1.00	16.00–18.00	...	...	0.15Si min
S43400	434	0.12	1.00	0.04	0.03	1.00	16.00–18.00	...	0.75–1.25	...
S43600	436	0.12	1.00	0.04	0.03	1.00	16.00–18.00	...	0.75–1.25	Nb:5xC-0
S44200	442	0.20	1.00	0.04	0.03	1.00	18.00–23.00	...	...	...
S44400	444	0.25	1.00	0.04	0.03	1.00	17.50–19.50	...	...	(Ti+Nb): 4(C+N)
S44600	446	0.20	1.50	0.04	0.03	1.00	23.00–27.00	...	...	0.25N
Martensitic grades										
S40300	403	0.15	1.00	0.04	0.03	0.50	11.50–13.00	...	...	...
S41000	410	0.15	1.00	0.04	0.03	1.00	11.50–13.50	...	...	...
S41400	414	0.15	1.00	0.04	0.03	1.00	11.50–13.50	1.25–2.50	...	...
S41600	416	0.15	1.25	0.06	0.15 min	1.00	12.00–14.00	...	0.60	...

UNS designation	AISI type	Composition, %(a) C	Mn	P	S	Si	Cr	Ni	Mo	Others
	416Se	0.15	1.25	0.06	0.06	1.00	12.00–14.00	. . .	. . .	0.15Se min
S42000	420	0.15 min	1.00	0.04	0.03	1.00	12.00–14.00	. . .	. . .	. . .
	420F	0.15 min	1.25	0.06	0.15 min	1.00	12.00–14.00	. . .	0.60	. . .
S42200	422	0.20–0.25	1.00	0.025	0.025	0.75	11.00–13.00	0.50–1.00	0.75–1.25	0.15–0.30 0.75–1.00
S43100	431	0.20	1.00	0.04	0.03	1.00	15.00–17.00	1.25–2.50	. . .	. . .
	440A	0.60–0.75	1.00	0.04	0.03	1.00	16.00–18.00	. . .	0.75	. . .
	440B	0.75–0.95	1.00	0.04	0.03	1.00	16.00–18.00	. . .	0.75	. . .
	440C	0.95–1.20	1.00	0.04	0.03	1.00	16.00–18.00	. . .	0.75	. . .
Precipitation-hardening grades										
S13800	S13800	0.05	0.20	0.010	0.008	0.10	12.25–13.25	7.50–8.50	2.00–2.50	0.90–1.35Al 0.01*N*
S15500	S15500	0.07	1.00	0.04	0.03	1.00	14.00–15.50	3.50–5.50		2.50-4.50 0.15–0
S17400	S17400	0.07	1.00	0.04	0.03	1.00	15.50–17.50	3.00–5.00		3.00–5.00cu 0.15–0
S17700	S17700	0.09	1.00	0.04	0.04	0.04	16.00–18.00	6.50–7.75		0.75–1.50Al

(a) Maximum unless otherwise indicated: all compositions include balance of iron.
(courtesy ASM International)

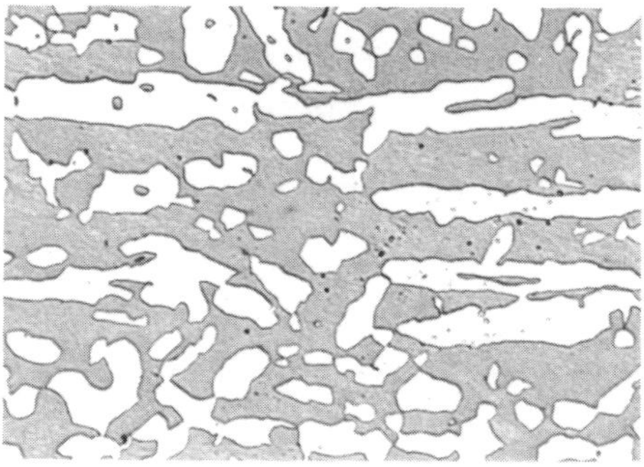

Figure 1–35 Microstructure of duplex stainless steel showing the austenite phase (light) and ferrite phase (dark). Magnified 400 times (courtesy NKK).

second, aging, causes the precipitation of a second phase that adds an additional strengthening increment.

The duplex stainless steels are not hardenable by heat treatment but generally are higher strength than the austenitics. The yield strength of annealed duplex stainless steels is generally about 70,000–80,000 psi. Moreover, duplex stainless steels have excellent corrosion resistance that is often better than either the ferrite or the austenite taken alone. The duplex stainless steels generally have a composition of 22–25% Cr, 5% Ni and 3% Mo. Sometimes nitrogen is added for enhanced pitting resistance. Figure 1–35 is an example of the duplex stainless steel microstructure.

There are many nonstandard stainless steels that are variations on these basic systems and are too numerous to mention. Some of those that have direct impact on the petroleum industry will be discussed in later chapters.

Nickel-Base Alloys

Just as in the case of stainless steels, there are a multitude of nickel-base alloys. The most common are the Incoloys, Inconels, Monels and Hastelloys. These alloys, the most common of which are presented in Table 1–4, are primarily composed of nickel, chromium, molybdenum, iron, aluminum, and titanium. Monels are nickel-copper alloys, Incoloys are nickel-iron-chromium alloys and Inconels are nickel-chromium alloys.

Many of the alloys are not heat-treatable and must be strengthened by cold

Table 1–4. Composition of Common Nickel-Base Alloys

UNS No.	Common Name	Composition, Wt. % (a) Cr	Mo	Ni	Fe	Co	Cu	Nb	W	Al	C	Ti	Si	Mn
N08028	Alloy 28	26-28	3.0-4.0	29.5-32.5	bal	-	0.6-1.4	-	-	-	0.03	-	1.00	2.50
N08825	Incoloy 825	19.5-23.5	2.5-3.5	38-46	bal	-	1.5-3.0	-	-	0.2	0.05	0.6-1.2	0.50	1.00
N07725	Incoloy 725	19.0-22.5	7.0-9.5	55-59	bal	-	-	2.75-4.0	-	0.35	0.03	1.0-1.7	0.20	0.35
N09925	Incoloy 925	19.5-23.5	2.5-3.5	38-46	22.0 min.	-	1.5-3.0	-	-	0.1-0.5	0.03	1.9-2.4	0.50	1.00
N07718	Inconel 718	17.0-21.0	2.8-3.3	50-55	bal	1.0	0.30	5.0	-	0.2-0.8	0.08	0.65-1.15	0.35	0.35
N07750	Inconel X-750	14.0-17.0	-	70 min.	5-9	-	0.50	0.7-1.2	-	0.4-1.0	0.08	2.25-2.75	0.50	1.00
N06625	Inconel 625	20-23	8.0-10.0	bal	5.0	-	-	-	-	0.40	0.10	0.40	0.50	0.50
N04400	Monel 400	-	-	63-70	2.5	-	bal	-	-	-	0.30	-	0.50	2.00
N05500	Monel K-500	-	-	63-70	2.0	-	bal	-	-	2.3-3.5	0.25	0.35-0.85	0.50	1.50
N07031	Pyromet 31	22.0-23.0	1.7-2.3	55-58	bal	-	0.6-1.2	-	-	1.0-1.7	0.03-0.06	2.1-2.6	0.20	0.20
N06975	Alloy 2550	23.0-26.0	5.0-7.0	47-52	bal	-	0.7-1.2	-	-	-	0.03	0.7-1.5	1.00	1.00
N06985	Hastelloy G-3	21.0-23.5	6.0-8.0	bal	18-21	5.0	1.5-2.5	0.5*	1.5	-	0.015	-	1.00	1.00
N06950	Hastelloy G-50	19.0-21.0	8-10	bal	16-18	2.5	0.5	-	1.0	-	0.02	-	1.00	1.00
N10276	Hastelloy C-276	14.5-16.5	15.0-17.0	bal	4.0-7.0	2.5	-	-	3.0-4.5	-	0.02	-	0.50	1.00
R30035**	MP35N	19.0-21.0	9.0-10.5	33-37	1.0	bal	-	-	-	-	0.025	1.0	0.15	0.15

(a) maximum unless otherwise indicated.
* Nb + Ta
** Actually Cobalt-base alloy

bal - balance
Incoloy and Inconel are trade names of INCO International.
Hastelloy is a trade name of Haynes.
Pyromet is a trade name of Carpenter.

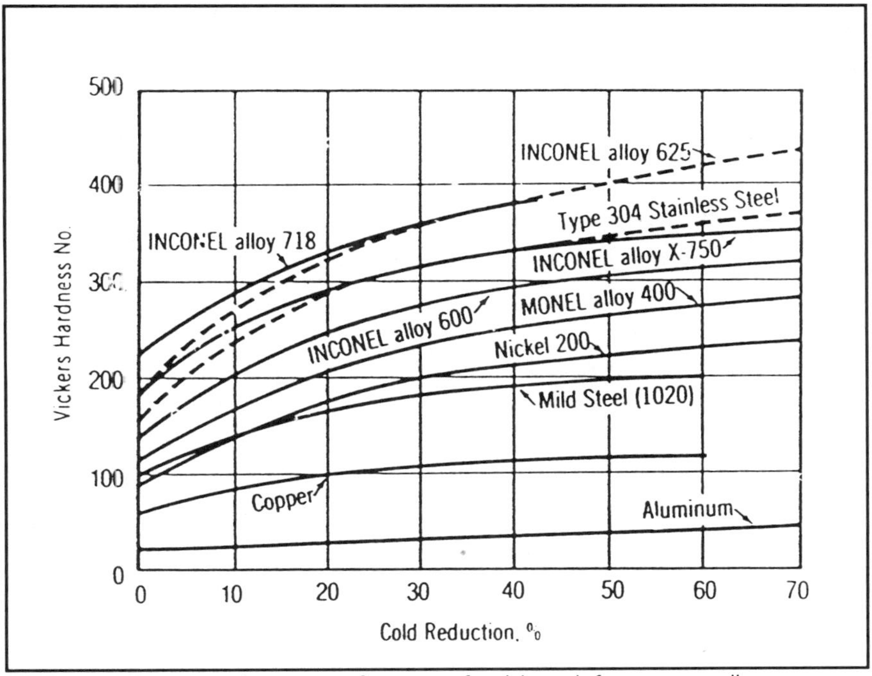

Figure 1–36 Hardness as a function of cold work for various alloys (courtesy Inco Alloys International, Inc.).

work. Other alloys can be strengthened by aging that produces a precipitate referred to as gamma prime, γ'.

Figure 1–36 shows the response of various alloys to cold work. The higher the degree of cold work, the greater the hardness up to a point after which the hardness is not measurably increased.

Figure 1–37 is an idealized presentation of aging that demonstrates the time-temperature relationship for hardening. As the temperature increases, the time to achieve maximum hardness is reduced. If the optimum time at temperature is exceeded, then overaging occurs and the precipitates enlarge reducing their strengthening effect. The γ' precipitate in a high-nickel alloy is most often composed of $Ni_3(Al, Ti)$ although niobium and tantalum may also produce γ'.

The nickel-base alloys are quite complex as a result of the numerous phases that can form. The most common phases are γ', eta (η), γ'', mu (μ), sigma (σ), laves and a variety of carbide and carbonitride phases. Some of these phases such as the chromium carbides are reversible and by proper heat treatment can be removed from the alloy. However, other phases such as sigma are practically irreversible and therefore present a serious problem if they are allowed to form. Many of these same phases also appear in stainless steels.

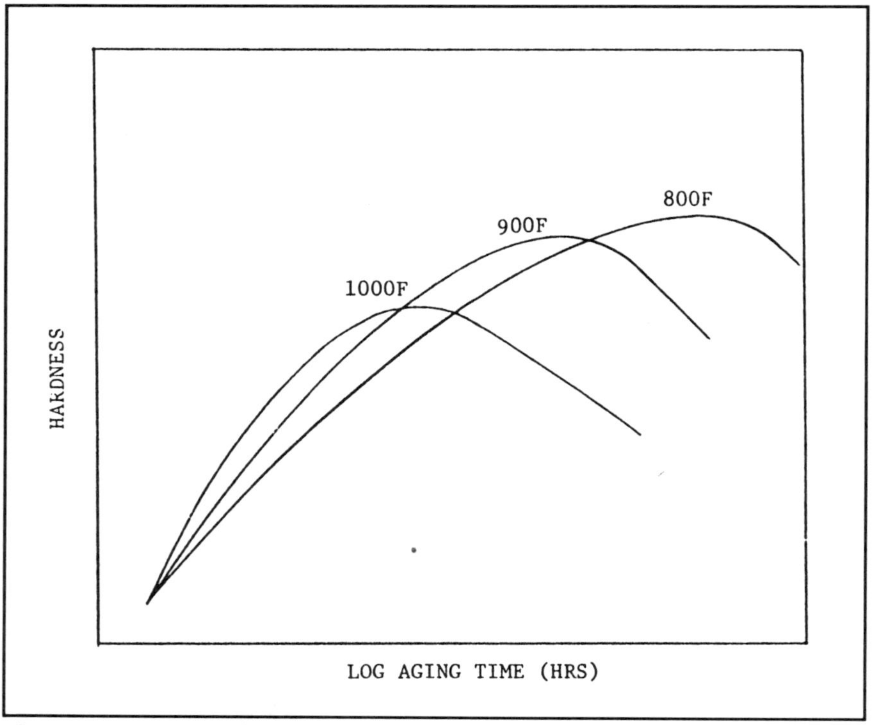

Figure 1–37 Hardness as a function of aging time and temperature for age-hardenable nickel alloys.

Titanium Alloys

Titanium is a very reactive metal that obtains its superior corrosion resistance from the stable titanium oxide (TiO_2) that forms on the surface.

Titanium has a higher temperature bcc phase referred to as beta and a low temperature (ambient) hcp phase called alpha. Titanium alloys are generally categorized as alpha, near alpha, alpha + beta, near-beta, and beta. Table 1–5 presents some of the more common Ti alloys. The major alloying elements are aluminum, nickel, vanadium, molybdenum, chromium, platinum, and tin. Aluminum and tin promote alpha phase stability while Ni, V, Mo, Cr, and Cu promote the beta phase.

Alpha titanium alloys cannot be strengthened by heat treatment. Likewise, near alpha alloys do not respond to heat treatment. Alpha + beta, near-beta, and beta alloys all display strengthening by heat treatment.

Alpha + beta alloys are generally liquid-quenched for the necessary rapid

Table 1–5. Some Common Titanium Alloys

UNS No.	Common Designation	Nominal composition, wt % Al	Sn	Zr	Mo	Others	Impurity limits, wt% N (max)	C (max)	H (max)	Fe (max)	O (max)
	Unalloyed grades										
—	ASTM Grade 1	...	...	...	...	...	0.03	0.10	0.015	0.20	0.18
R50400	ASTM Grade 2	...	...	...	...	...	0.03	0.10	0.015	0.30	0.25
—	ASTM Grade 3	...	...	...	...	...	0.05	0.10	0.015	0.30	0.35
—	ASTM Grade 4	...	...	...	...	...	0.05	0.10	0.015	0.50	0.40
—	ASTM Grade 7	...	...	...	...	0.2Pd	0.03	0.10	0.015	0.30	0.25
	Alpha and near-alpha alloys										
R53400	TiCode 12	...	...	...	0.3	0.8Ni	0.03	0.10	0.015	0.30	0.25
—	Ti-5Al-2.5Sn	5	2.5	...	...	...	0.05	0.08	0.02	0.50	0.20
—	Ti-5Al-2.5Sn-ELI	5	2.5	...	...	...	0.07	0.08	0.0125	0.25	0.12
—	Ti-8Al-1Mo-1V	8	...	...	1	1V	0.05	0.08	0.015	0.30	0.12
—	Ti-6Al-2Sn-4Zr-2Mo	6	2	4	2	...	0.05	0.05	0.0125	0.25	0.15
—	Ti-6Al-2Nb-1Ta-0.8Mo	6	...	...	1	2Nb,1Ta	0.02	0.03	0.0125	0.12	0.10
—	Ti-2.25Al-11Sn-5Zr-1Mo	2.25	11.0	5.0	1.0	0.2Si	0.04	0.04	0.008	0.12	0.17
	Alpha-beta alloys										
—	Ti-6Al-4V	6.0	...	...	...	4.0V	0.05	0.10	0.0125	0.30	0.20
—	Ti-6Al-4V-ELI	6.0	...	...	...	4.0V	0.05	0.08	0.0125	0.25	0.13
—	Ti-6Al-6V-2Sn	6.0	2.0	...	...	0.75Cu,6.0V	0.04	0.05	0.015	1.0	0.20
—	Ti-8Mn	...	...	...	...	8.0Mn	0.05	0.08	0.015	0.50	0.20
—	Ti-7Al-4Mo	7.0	...	...	4.0	...	0.05	0.10	0.013	0.30	0.20
R56260	Ti-6Al-2Sn-4Zr-6Mo	6.0	2.0	4.0	6.0	...	0.04	0.04	0.0125	0.15	0.15
—	Ti-5Al-2Sn-2Zr-4Mo-4Cr	5.0	2.0	2.0	4.0	4.0Cr	0.04	0.05	0.0125	0.30	0.13
—	Ti-10V-2Fe-3Al	3.0	...	...	...	10.0V	0.05	0.05	0.015	2.5	0.16
—	Ti-3Al-2.5V	3.0	...	...	...	2.5V	0.015	0.05	0.015	0.30	0.12
	Beta alloys										
—	Ti-13V-11Cr-3Al	3.0	...	...	...	11.0Cr,13.0V	0.05	0.05	0.025	0.35	0.17
—	Ti-8Mo-8V-2Fe-3Al	3.0	...	...	8.0	8.0V	0.05	0.05	0.015	2.5	0.17
R58640	Ti-3Al-8V-6Cr-4Mo-4Zr (Beta C)	3.0	...	4.0	4.0	6.0Cr,8.0V	0.03	0.05	0.020	0.25	0.12
—	Ti-11.5Mo-6Zr-4.5Sn (Beta III)	...	4.5	6.0	11.5	...	0.05	0.10	0.020	0.35	0.18

Figure 1–38 Microstructure of Ti–6Al–4V magnified 100 times. The white phase is alpha and the dark phase beta.

cooling to provide optimum response to aging. However, beta alloys are typically air- quenched from the solution annealing temperature. Both types of alloys are then aged and strengthening occurs by the decomposition of the supersaturated beta phase. Figure 1–38 shows the mixed alpha and beta phase in Ti–6Al–4V.

One of the advantages of titanium alloys over nickel-base alloys is the high strength that can be achieved by heat treatment. For example, low-alloy steels and titanium alloys can be heat treated to produce useable yield strengths as high as 180,000 psi. However, age-hardenable nickel alloys are limited to about 130,000 psi.

References

American Petroleum Institute. *Recommended Practice 2A.*

American Society for Metals. "Properties and Selection of Metals." In *Metals Handbook.* Volume 1. 1961, pp. 227 and 236; 1978, p.494, Volume 13, 1987.

American Society for Testing and Materials. *Standards A370* and *E23.*

Bravenec, E.V. "Analysis of Brittle Fractures During Fabrication and Testing." *ASME Pressure Vessel and Piping Division Meeting.* Paper 53. San Francisco, California, May 1981.

Cias, W.W. *Phase Transformation Kinetics and Hardenability of Medium-Carbon Alloy Steels.* Greenwich, Connecticut: Climax Molybdenum Company.

Dieter, G.E. Jr. *Mechanical Metallurgy.* New York, New York; McGraw-Hill Book Company, 1961.

Databook. *Metal Progress.* Volume 122. 1982, p. 75.

Honeycombe, R.W.K., *Steels-Microstructure and Properties,* ASM International, 1982.

Huntington Alloys Handbook, *Fabricating,* p. 3, 1981.

Reed-Hill, R.E. *Physical Metallurgy Principles.* New Jersey: D. Van Nostrand Company Inc. 1964.

Stainless – 12%+ Cr

Duplex – 22-25% Cr
5% Ni
3% Mo

Nickel Base – Inconel 50% Ni

2

FUNDAMENTALS OF CORROSION

With few exceptions, metals and alloys are placed in a service environment that in some way alters or degrades the material. This degradation is referred to as *corrosion*, and no study of metallurgy can be complete without discussing the effect of environments on metals. These effects can manifest themselves in many ways, which will be described in this chapter. Furthermore, corrosion is a frequent downfall of a well-designed alloy system that has been carefully planned based on physical and mechanical properties but whose performance has never been considered in an aggressive environment. Since petroleum production is by its very nature corrosive, the remainder of this book will deal often with the effect of corrosion on metals.

The Corrosion Cell

Corrosion is primarily an electrochemical process, which means that in addition to the chemical reaction there is also the movement of electrons and, as such, a current flow. Since current must flow from one location to another, there are two basic reactions occurring. One involves the discharge of electrons; the surface that supports this reaction is called the anode. The other reaction, which consumes the electrons, is called the cathode. Figure 2–1 represents the basic corrosion cell that includes an anode where metal ions go into solution, and a cathode where electrons are consumed. Electrons move along the path of low resistance.

The anode reaction is called *oxidation* and is represented as metal (M) going to metal ions (M^+):

$$M \rightarrow M^+ + e$$

Likewise, to conserve electrons (e) as required, the following reaction occurs in acid solutions at the cathode and is referred to as *reduction:*

$$2H^+ + 2e \rightarrow H_2$$

If a voltmeter were introduced into the circuit between the two metal strips, a steady-state potential or voltage would be observed after a short equilibration time. The magnitude of this potential is a function of the type of metal, the solution chemistry, and the temperature. The concept of potential is impor-

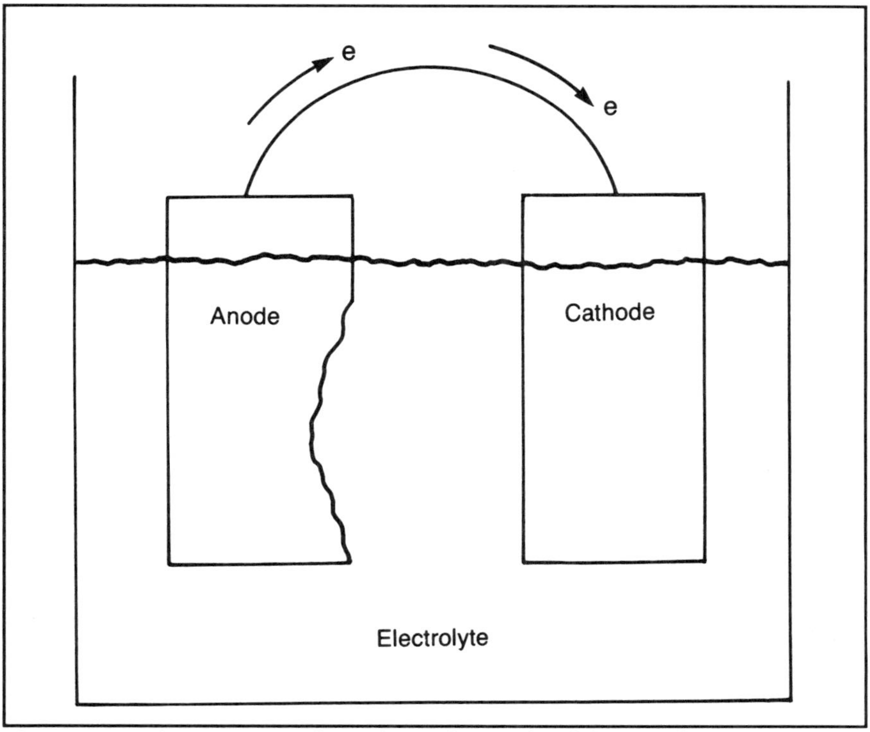

Figure 2–1 Components of an electrochemical cell that produce corrosion.

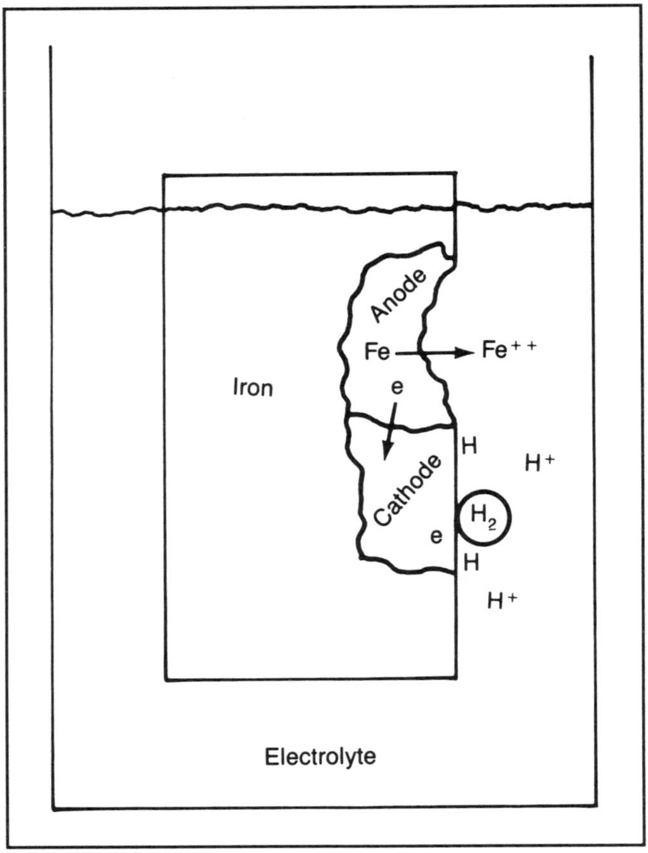

Figure 2–2 Schematic of local anodes and cathodes on steel surface.

tant since, as will be demonstrated in later examples, the potential of a metal in solution is an indication whether it will corrode or not.

There are a multitude of oxidation/reduction reactions, but the more common ones in the oilfield are:

Oxidation:	$Fe \rightarrow Fe^{+2} + 2e$
	$Fe^{+2} \rightarrow Fe^{+3} + e$
Reduction:	$2H + 2e \rightarrow H_2$ (acidic solutions)
	$O_2 + 4H^+ + 4e \rightarrow 2H_2O$ (acid solutions)
	$O_2 + 2H_2O + 4e \rightarrow 4OH^-$ (neutral or basic solutions)

It is conventional to add the two reactions to describe the complete electrochemical reaction such that iron corroding in hydrochloric acid would be written:

$$Fe + 2HCl \rightarrow FeCl_2 + H_2$$

Anodes and cathodes are not always two separate metals but in fact are commonly spread out over the surface of one metal. This is the result of small variations in composition, surface roughness, cold work, and surface films, to name the primary ones. Figure 2–2 represents the corrosion of steel in an acid solution. Here, the easy electron path is within the metal itself and is not part of an external connection.

Considering the reaction just described for steel in an acid solution, certain characteristics of the environment affect the corrosion reaction. Factors that increase either the anodic or the cathodic reaction increase the corrosion rate since the two are linked. The principal factors are oxygen or oxidizer concentration, temperature, velocity, and concentration of reactants.

None of these parameters has a simple relationship to corrosion rate. For instance, increasing temperature is detrimental up to the boiling point because it increases the reaction rate. A common rule of thumb is, for every 10°C increase in temperature, the reaction rate (corrosion rate) doubles. However, above the boiling point, the corrosion rate may actually decrease as the corrosion mechanism changes to a high-temperature attack.

For velocity, the same phenomenon occurs when initially increasing velocity increases the corrosion rate, but at some point the corrosion rate may decrease since reactants do not have sufficient time to initiate corrosion at the metal surface. However, for many applications, increasing velocity, temperature, oxygen, and concentration of corrodents generally can be considered to increase the corrosion rate.

In petroleum production and drilling, the primary agents that cause corrosion are oxygen (O_2), carbon dioxide (CO_2), hydrogen sulfide (H_2S), and chlorides. More recently, elemental sulfur (S) produced from gas and oil wells has been encountered, and it, too, is corrosive. Besides these naturally occurring corrodents, there is the action of hydrochloric (HCl) and hydrofluoric (HF) acids, both of which are used in acid stimulation and the former in wellbore cleanup. Likewise, clear brines for completion and packer fluids can be quite corrosive. The advent of tertiary oil recovery has brought numerous other corrodents into the picture. The specific role of each of these constituents will be discussed in subsequent chapters.

The concept of corrosion rate and the principles behind it have been discussed, but how is it actually measured? One method is to measure the amount of metal loss by weight change. This approach is not satisfactory since sample geometry and time of exposure can affect the amount of weight lost, making comparison with other methods inaccurate. In fact, the best expressions for corrosion rate use penetration per unit time in units of inches per year (ipy), millimeters per year (mm/y), microns/year (μm/y) or, mils per year (mpy). A conversation table for these common rates is presented in Table 2–1.

TABLE 2–1. Conversions for Corrosion Rate Units

Unit *	Factor mpy	ipy	mm/y	μm/y
Mils/year (mpy)	1	0.001	0.0254	25.4
inch/year (ipy)	1000	1	25.4	25,400
Millimeter/year (mm/y)	39.4	0.0394	1	1000
microns/year (μm/y)	0.0394	0.0000394	0.001	1

* Multiply the unit in this column by the appropriate factor to obtain the new unit in the right columns.

A simple relationship for determining corrosion rate in mils per year from weight-loss data is (from Fontana and Greene):

$$\text{mpy} = \frac{534\ W}{DAT}$$

Where:

W = weight loss, mg
D = density of specimen, g/cm^3
A = area of specimen exposed, sq in.
T = exposure time, hr

Techniques for measuring corrosion rates vary considerably, from the simple immersion of preweighted strips of metal called coupons to the more complex potentiometric methods. While these methods are important, they are not germane to this book. Rather, it is more appropriate here to present the various forms of corrosion as they relate to the oilfield. These forms will be referred to throughout the book to explain the reasons for metallurgical choices in drilling, production, and handling of oil and gas.

Types of Corrosion

Although corrosion has been categorized various ways by many authors, there is no one universal system for classifying the forms of corrosion. However, it is convenient to organize the types as uniform, pitting, crevice, dealloying, intergranular, galvanic, erosion/impingement, and environmentally assisted cracking.

Uniform corrosion represents the ideal case in which the metal is uniformly corroded away at some constant rate. This particular form of attack—while rarely seen in practice—is the basis for most design. Using a predetermined corrosion rate of so many mils per year, the designer assumes corrosion will progress in a uniform manner. In fact, frequently corrosion is very localized, and failure occurs most often by this means long before failure by general thinning of the metal. Uniform or general corrosion is commonly observed on metal structures exposed to the atmosphere (e.g., offshore platforms, exterior of tanks, etc.).

Pitting attack is one of the most frequent forms of corrosion encountered. It represents a very localized attack and may cause failure in a matter of weeks or months while the remaining area of metal is relatively uncorroded. Figure 2–3 shows an example of pitting attack on a section of tubing that was used in sour (H_2S) oil production. The shape of pits varies with the particular pitting environment. The pits shown in Figure 2–3 are the result of corrosion by hydrogen sulfide and water in the produced fluids.

It is generally accepted that once pitting is initiated, the remaining unpitted surface area becomes cathodic to the pits. As the pits progress, the pH of the solution in the bottom of the pit is reduced, often to a pH of 1. This pH

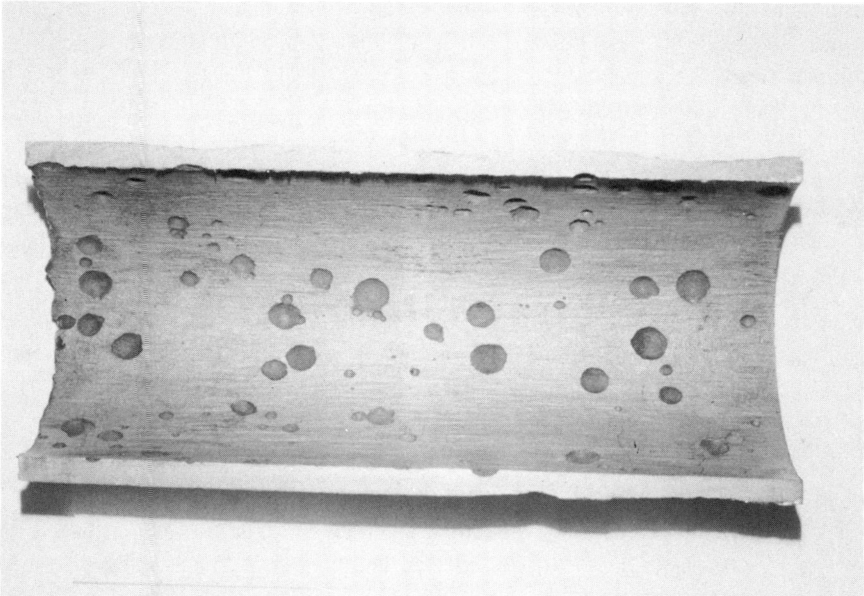

Figure 2–3 Pitting corrosion in J-55 tubing from crude oil containing H_2S.

reduction, along with the anodic character of the pit, drives the corrosion process until the metal is perforated or a change in the environment stifles the continuing reaction. At any given time during the pitting of a metal surface, a statistical distribution of pit depths will range from almost nil to a maximum depth. Most pits will have some average depth between these two extremes. During the time it takes to perforate the metal, some pits initiate and then stop growing, while others propagate to various depths and stop or continue to grow until failure. Currently, there is not a good understanding as to why some pits grow to completion while others may initiate, propagate, and arrest. This explains the difficulty in predicting pitting corrosion and its rate of attack. Moreover, it is dangerous to attempt to estimate the remaining life of a structure based on pit depth measurements. While pitting can be catastrophic and difficult to predict, much is known about which environments induce pitting for many alloy systems.

The chloride ion is a particularly aggressive pitting agent for low-alloy and stainless steels, as shown in Table 2–2. Chloride pitting typically initiates at inclusions (usually sulfide inclusions) in the steel because the local alloy chemistry is significantly different than in the rest of the matrix. Carbon and low-alloy steels also pit in solutions containing hydrogen sulfide, carbon dioxide, and oxygen. Dissolved oxygen enhances pitting corrosion, especially in the presence of hydrogen sulfide, carbon dioxide, and chlorides.

Figure 2–4 shows the corrosion rate of steel as a function of dissolved oxygen content in water. Corrosion of steel from dissolved oxygen is almost always by pitting. Note how little oxygen is necessary, in the parts per million (ppm) range, to cause attack. It has been found that total elimination of oxygen in oilfield waters is almost impossible but a reasonable threshold below which attack does not occur is 20 parts per billion (ppb).

Carbon dioxide and hydrogen sulfide also produce pitting and while under some circumstances they may lead to uniform corrosion, it is more common to observe pitting attack. Unlike oxygen, these gases generally cause corrosion when present in concentrations on the order of tenths of a percent or greater.

Oxygen also aids attack of another form of corrosion—crevice corrosion. This type of attack arises from two surfaces in contact with each other. These contacts can be gaskets, metal touching metal, corrosion products, or deposits that have settled on the metal surface.

Deposits can result by the precipitation of mineral scales ($CaSO_4$, $CaCO_3$, or $BaSO_4$) from water, or sand and silt lying at the bottom of a line. The moisture under the deposit is depleted in oxygen compared to the rest of the solution. This variation in oxygen concentration will affect the potential of the metal as described earlier so that with two different potentials a corrosion cell is set up. The oxygen-rich surfaces of the metal act as cathodes, while the metal under the

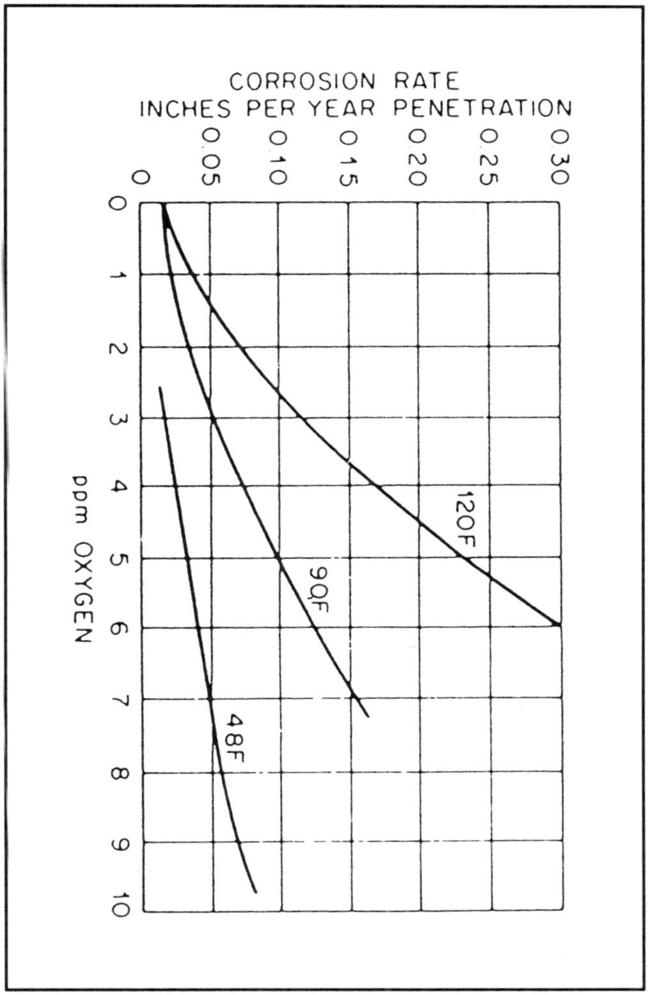

Figure 2-4 Corrosion rate of steel as a function of dissolved oxygen content (courtesy Betz Laboratories).

deposit becomes the anode. Because of the importance of oxygen in this type of attack, it is also referred to as oxygen depletion cell corrosion or concentration cell corrosion.

However, there is a critical size range in which crevice corrosion can occur. If the opening is too narrow, the solution will not easily enter the crevice; and if it is too large, mixing with the bulk solution will not allow the development of a stagnant region in the crevice. In general, a crevice of approximately 0.001–0.050 in. is necessary, although the particular situation will affect this size range.

Table 2–2. Minimum Concentration of Cl^- Necessary for Pit Initiation in 1N H_2SO_4 Solution.

Alloy	Cl^- (Normality)
Fe	0.0003
Fe-5.6 Cr	0.017
Fe-11.6 Cr	0.069
18.6 Cr–9.9 Ni–Fe	0.1
20.0 Cr–Fe	0.1
24.5 Cr–Fe	1.0
29.4 Cr–Fe	1.0

(After Stolica, reprinted with permission from Pergamon Press Ltd.)

Dealloying or selective leaching is the selective corrosion of one element in an alloy. Dezincification is the classic example of dealloying whereby zinc is selectively attacked and removed from brass alloys, leaving only the copper behind. Gray cast iron also suffers this type of attack in which the iron matrix is selectively corroded, leaving behind a layer of graphite. This attack of cast irons has been called graphitization or graphitic corrosion. Gray cast-iron butterfly valves and water pumps used in the petroleum industry are most susceptible to this type of attack. Table 2–3 gives a brief description of some of the alloys susceptible to selective leaching and the element removed.

Intergranular corrosion occurs most frequently in austenitic stainless steels as a result of heat treatment or welding that promotes a change in chemistry between the grain boundaries and the grains. These treatments leave the regions adjacent to the grain boundaries depleted in chromium compared to the metal matrix leading to a higher degree of susceptibility to corrosion. Corrosion then proceeds along the grain boundary network at a much higher rate than in the matrix. Figure 2–5 shows an example of intergranular corrosion of a CF8M (316) stainless steel pump impeller. The critical temperature change for inducing susceptibility to this form of attack is 950–1,450°F regardless if the alloy is held in this temperature range or slowly cooled through it. Thus, it is common practice to rapidly quench the austenitic stainless steels from above 1,800°F to avoid the problem. Low-carbon or stabilized grades of austenitic stainless steels are used to avoid this problem.

Galvanic corrosion is one of the most common yet least-recognized corrosion problems. When two or more dissimilar metals are connected electrically and placed in a solution, current will flow between them because of the potential difference of the metals. The metal with least resistance to corrosion (active metal) in the particular environment will become the anode, while the more corrosion-resistant metal (noble metal) will become the cathode. The two metals do not have to be connected by standard electrical means; intimate contact will

Table 2-3. Combinations of Alloys and Environments Subject to Selective Leaching, and Element Removed by Leaching.

Alloy	Environment	Element Removed
Brasses	Many waters, especially under stagnant conditions	Zinc (dezincification)
Gray iron	Soils, many waters	Iron (graphitic corrosion)
Aluminum bronzes	Hydrofluoric acid, acids containing chloride ions	Aluminum
Silicon bronzes	Not reported	Silicon
Copper nickels	High heat flux and low water velocity (in refinery condenser tubes)	Nickel
Monels	Hydrofluoric and other acids	Copper in some acids, nickel in others
Alloys of gold or platinum with nickel, copper, or silver	Nitric, chromic, and sulfuric acids	Nickel, copper, or silver (parting)
High-nickel alloys	Molten salts	Chromium, iron, molybdenum, and tungsten
Cobalt-tungsten-chromium alloys	Not reported	Cobalt
Medium-carbon and high-carbon steels	Oxidizing atmospheres, hydrogen at high temperatures	Carbon (decarburization)
Iron-chromium alloys	High-temperature oxidizing atmospheres	Chromium, which forms a protective film
Nickel-molybdenum alloys	Oxygen at high temperature	Molybdenum

(After *Metals Handbook*, Volume 10)

suffice. Corrosion of the anode will usually be more severe than corrosion of that metal alone in the same solution, while the cathode will achieve a degree of protection from the environment—sometimes to the extent that corrosion is completely stopped on the cathodic metal. These effects can be measured and

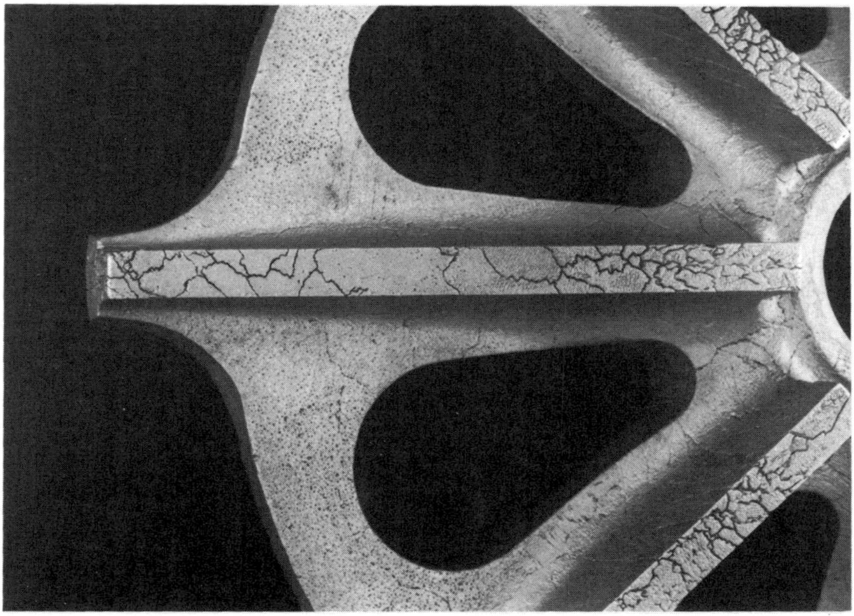

Figure 2–5 Intergranular corrosion of an austenitic stainless steel pump impeller.

have been done for couples of metals in seawater at 25°C (see Table 2–4). Table 2–4 provides a relative ranking of metals and alloys regarding their resistance to corrosion in seawater. The greater the distance between two metals in the table, the greater their potential difference and the higher the probability the active metal will suffer accelerated corrosion.

Note that some alloys and metals are listed twice in the table—once with the word "active" following and once with the word "passive." Some metals and alloys become essentially immune to corrosion in certain environments because of the formation of a surface film so thin that it is impossible to see with the naked eye or even with an optical microscope. The stability of these films is paramount to the enhanced corrosion resistance of these alloys. Moreover, corrosion films represent the controlling factor in almost all corrosion (see Craig). Although little is known about the actual construction of these passive films, considerable knowledge exists concerning the effect of alloying on the formation of these films in different alloy systems. Thus the addition of chromium to iron can produce a passive alloy (18-8 stainless steel) of considerable corrosion resistance compared to the original iron. However, in an environment where the passive film is not functional, the active surface becomes far less noble as indicated in Table 2–4.

A more active metal in the series will corrode at the expense of a more noble one. Thus, coupling zinc to steel will cause the zinc to corrode and will protect the steel. This is the reason for galvanizing steel; when pinholes in the galvanizing occur, the steel underneath, once exposed to the environment, will be protected by the zinc. This is also the basis for cathodic protection. Sacrificial anodes are made of metals or alloys that are more active than steel, which allows for the consumption of the anode and the protection of the steel structure. However, if steel is coupled to copper, the distance on the chart is large so steel in this case will be the anode and have a greater tendency to corrode.

The galvanic series is useful for approximating the behavior of coupled alloys; however, care must be used in its application. Several parameters affect galvanic corrosion and as such may affect the actual behavior of a couple in service. Two important factors in galvanic corrosion are the temperature and the relative area of the metals. Increasing temperature in many cases may cause a reversal in the anode/cathode relationship. The galvanic series is applicable to many oilfield waters at ambient temperature; however, the zinc/steel couple has been found to reverse at about 140°F such that zinc becomes the cathode. This reversal has been responsible for failures of galvanized systems or systems protected with zinc sacrificial anodes. It has also been qualitatively determined that as the carbonate concentration of the solution increases, the temperature of potential reversal for the zinc/iron system decreases. These effects point to the need to measure the potential of a couple, especially in cathodic protection, in the actual environment prior to its application. It must always be borne in mind that Table 2–4 relates only to seawater and that extending it to other environments may result in some changes in the position of metals and alloys in the series.

The other factor, area-of-anode-to-area-of-cathode, is of considerable importance. If the anode area is smaller than the cathode area, the corrosion rate may be increased many orders of magnitude as a function of this ratio. However, if the anode area is greater than the cathode area, corrosion of the anode will be less than for a 1:1 anode/cathode ratio.

This behavior is shown in Figure 2–6. In this figure, the steel pipe pump column was coated with an epoxy coating and flanged to an aluminum-bronze seawater pump. Since the coating had several small pinholes (holidays) in it, the area of the steel anodes was small in comparison to the large, bronze pump cathode area. Penetration occurred in a matter of weeks. An important but difficult concept to understand intuitively from this example is that, when a couple such as steel to bronze must be made, the cathode (bronze) should be coated to reduce its area in relation to the anode.

Galvanic corrosion can also occur when metals of the same composition are coupled. If a section of pipe is removed from a pipeline that has been in service

Figure 2–6 Galvanic corrosion of seawater pump casing. The steel casing was coated with an epoxy, substantially decreasing the anode area. Worst attack is in the flange area where the casing was flanged to a bronze pump.

for many years and a new piece is installed, the new section will be anodic to the remaining line since the older sections have had time to form corrosion-product films. Generally, this is not a problem since the new pipe will corrode quickly at first while it also forms a comparable corrosion film. Another type of galvanic corrosion occurs most often in wet-gas systems containing carbon dioxide. Here, the heat-affected zone of a weldment is aggressively corroded while the weld bead is unaffected. This attack is the result of a high residual stress inherent in the heat-affected zone caused by welding in comparison to the lower stress in the weld bead, which experienced a slower cooling rate. It is also attributed to the metallurgical structure of the weld. However, depending on the chemistry of the welding rod, this galvanic cell can reverse and the weld bead can become anodic to the heat-affected zone.

Erosion-corrosion and impingement represent another form of corrosion. As discussed at the beginning of this chapter, velocity effects are quite complex and it is dangerous to make general statements concerning them; however, for most cases velocity increases corrosion, especially in the presence of fine particles such as sand or silt. Erosion-corrosion involves both the velocity contribution to corrosion as well as the mechanical degradation caused by the abrasion from

Table 2–4. Galvanic Series of Some Commercial Metals and Alloys in Seawater

Active or anodic	Magnesium
	Magnesium alloys
	Zinc
	Galvanized steel
	Aluminum 1100
	Aluminum 2024 (4.5 Cu, 1.5 Mg, 0.6 Mn)
	Mild steel
	Wrought iron
	Cast iron
	13% Chromium stainless steel
	Type 410 (active)
	18—8 Stainless steel
	Type 304 (active)
	Lead-tin solders
	Lead
	Tin
	Muntz metal
	Manganese bronze
	Naval brass
	Nickel (active)
	76 Ni-16 Cr-7 Fe alloy (active)
	60 Ni-30 Mo-6 Fe-1 Mn
	Yellow brass
	Admiralty brass
	Red brass
	Copper
	Silicon bronze
	70-30 Cupro nickel
	G-bronze
	Silver solder
	Nickel (passive)
	76 Ni-16 Cr-7 Fe (Inconel alloy) (passive)
	13% Chromium stainless steel
	Type 410 (passive)
	Titanium
	18-8 Stainless steel
	Type 304 (passive)
	Silver
Noble or cathodic	Graphite
	Gold
	Platinum

(After NACE)

these fine particles. Impingement is usually considered the more direct blasting of the surface by the fluid. Solids and gas bubbles have a tendency to increase the wastage caused by impingement.

Figure 2–7 Impingement attack of piping caused by high-velocity gas flow.

Figure 2–7 shows an example of impingement in which a wet gas stream flowing from a pipe entered another pipe at 90°, causing impingement and leaving a highly polished, localized penetration in the wall.

Erosion-corrosion is actually a synergistic effect that is more than the sum of the components. Increasing velocity resulting in erosion causes deterioration by wear; but with the presence of corrodents, the surface is degraded faster than

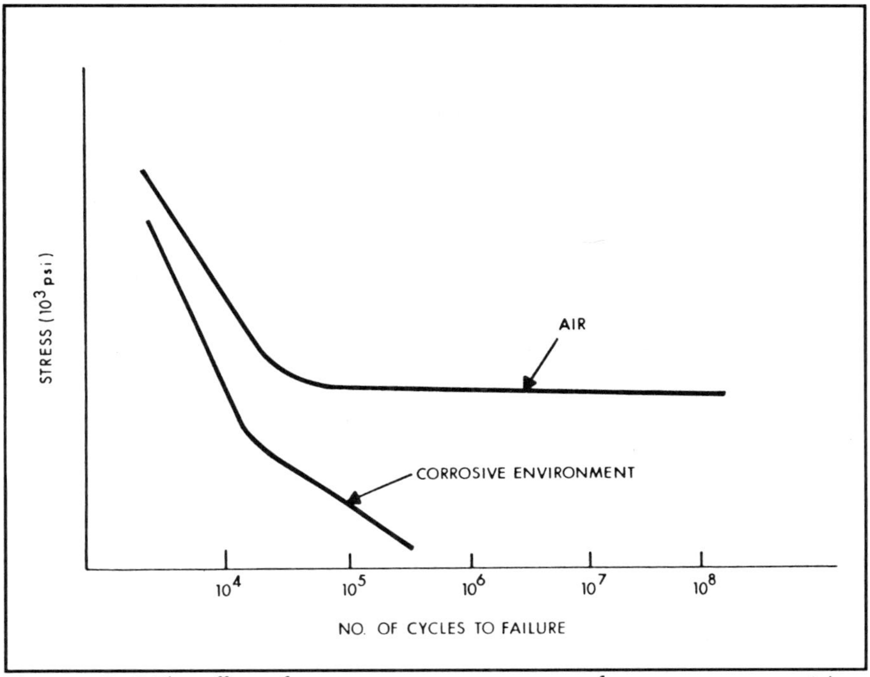

Figure 2–8 The effect of a corrosive environment on fatigue resistance. Note the absence of adefinite fatigue limit in the corrosive environment.

that expected from erosion or corrosion alone. In fact, in many forms of corrosion, there is a synergistic effect between some mechanical action that removes the corrosion product layer and corrosion that attacks the fresh metal.

Probably one of the most catastrophic forms of corrosion is environmentally assisted cracking. There are three basic categories that fall into this type of corrosion cracking: corrosion fatigue, stress-corrosion cracking, and hydrogen damage.

The first of these involves a corrosive environment superimposed on the purely mechanical nature of fatigue. Some aspects of fatigue are discussed in the first chapter, and Figure 1–18 shows the fatigue behavior of steel in a noncorrosive environment. Figure 1–19 of the same chapter shows the deleterious effect of a saltwater environment on fatigue resistance, demonstrating the complete loss of a threshold stress or fatigue limit. In many cases, the loss of a threshold is not as abrupt and the corrosion fatigue curve appears more like that shown in Figure 2–8.

Fracture most often initiates at the surface of the metal after most of the fatigue life has been spent prior to crack initiation. Once a crack is formed,

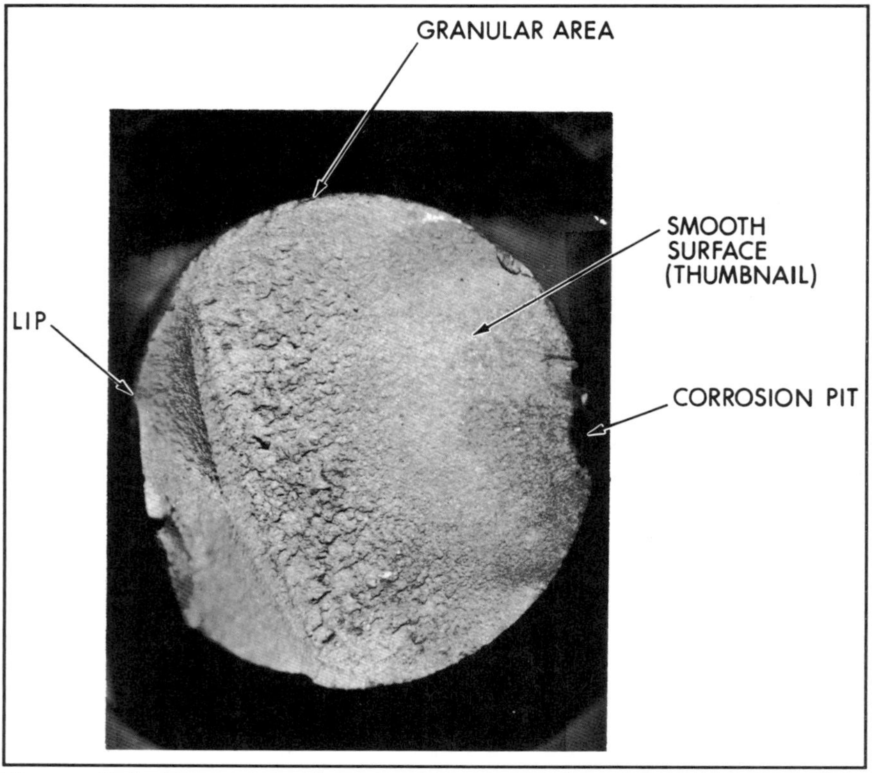

Figure 2–9 Characteristic features of a corrosion-fatigue fracture on a sucker rod.

complete fracture, which is propagation to failure, occurs in about 20% of the remaining life of the part. The fatigue fracture surface is quite distinctive, allowing easy recognition of this type of failure. Figure 2–9 shows the typical corrosion-fatigue failure of an oilfield sucker rod. A pit on the right has initiated the crack that propagated perpendicular to the tensile stress on the rod string. The smooth region represents stable crack growth and has the characteristic thumbnail appearance of a crack advancing from a single initiation point. In many fatigue failures, the faces do not rub each other, and the series of concentric rings (beach marks) are evident as shown in Figure 1–20. Once crack growth has progressed to the point that the remaining cross section cannot sustain the applied load, rapid, unstable fracture occurs, often leaving a bright crystalline appearance and a 45° shear lip. In ferrous metals, the corrosion-fatigue fracture is usually transgranular with very little branching present.

Unlike purely mechanical fatigue, corrosion fatigue is a strong function of the stress cycle frequency. If cycling is quite rapid, sufficient time is not allowed for

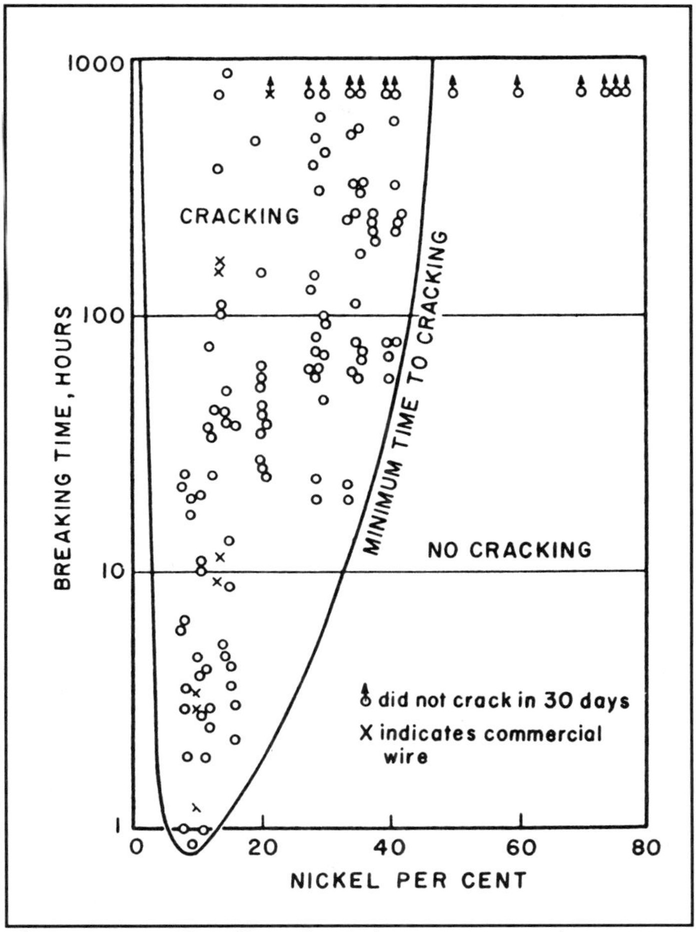

Figure 2-10 The effect of nickel content in iron-chromium-nickel stainless steel on susceptibility to stress-corrosion cracking in boiling Mg_2Cl (After Copson).

contact between the crack tip and the corrosive environment to provide a contribution from the corrodents. On the other hand, very slow stress cycling can allow corrosion to blunt the crack tip, arresting or slowing the mechanical action of fatigue. Thus, there is a range of cyclic stress that is particularly conducive to corrosion fatigue.

Another form of cracking is stress-corrosion cracking, which is caused by the combined action of corrosion and tensile stress. Stress-corrosion cracking (SCC) susceptibility for an alloy is a function of solution chemistry, alloy composition, microstructure, stress, pH, specific ion concentration, and temperature. Not

Table 2-5. Evaluating Media That Causes Stress-Corrosion Cracking

● indicates media is a corrodent for given material

Corrosive Media	1	2	3	4	5	6	7	8	9	10	11	12	13	14	15	16	17	18	19	20	21	22
Acetic acid + mercury salts								●	●	●								●				
Alkyl aryl sulfonates				●																		
Aluminum chloride				●	●																	
Aluminum sulfate				●																		
Ammonia			●						●	●												
Ammonium biphosphate				●	●																	
Ammonium chloride				●	●																	
Ammonium fluosilicate													●									
Ammonium hydroxide								●	●	●												
Ammonium nitrate			●																			
Aniline										●												
Barium chloride				●	●																	
Barium nitrate			●																			
Bismuth								●										●				
Calcium bromide				●	●																	
Calcium chloride					●																	
Calcium nitrate			●																			
Carbon tetrachloride				●	●																	
Chlorine				●	●																	
Chloranilines										●												
Chlorobenzene													●									
Chloroform				●	●																	
Chlorotoluidines										●												
Chromic acid			●							●			●									
Copper tetramine										●												
Cresol			●																			
Cresylic acid vapors			●																			
Cyanogen			●																			
Dichlorophenol				●	●																	
Epichlorhydrin				●	●																	
Ethanol																					●	
Ethylamine			●																			
Ethyl chloride				●	●																	
Ferric chloride			●							●												●
Ferrous chloride				●	●																	
Fluosilicic acid												●	●			●	●					
Gasoline vapor				●																		
Glutamic acid				●	●																	
Glycerol				●																		
Hexachloroethane				●	●																	
Hydrochloric acid (aerated)																		●				
Hydrocyanic acid + hydrogen cyanide				●		●																
Hydrofluoric acid vapors													●									
Hydrofluoric acid (aerated)				●									●									
Hydrofluoric acid (no air)									●	●		●	●	●								
Hydrogen				●																		
Hydrogen chloride			●	●																		
Hydrogen sulfide (wet)				●														●				
Lead acetate																				●		
Lead bromide				●																		
Levulinic acid				●	●																	
Lithium			●						●	●												
Lithium chloride				●	●																	
Magnesium chloride			●	●	●	●			●	●			●									
Magnesium chloride + calcium chloride				●																		

Corrosive Media	1	2	3	4	5	6	7	8	9	10	11	12	13	14	15	16	17	18	19	20	21	22
Magnesium fluosilicate												●	●									
Magnesium sulfate																●	●					
Manganese chloride				●	●																	
Mercuric chloride				●	●			●	●	●	●		●					●				
Mercuric cyanide													●									
Mercuric nitrate													●									
Mercurous nitrate				●				●	●	●			●									
Mercury				●					●	●		●	●	●				●				
Methallylamine										●												
Methanol																					●	
Monoethanolamine			●																			
Naphtha feedstock				●																		
Naphthenic acids				●																		
Nickel chloride				●																		
Nickel nitrate			●																			
Nitric acid			●							●												
Pentachloroethane				●	●																	
Perchloroethylene																					●	
Phosphoric acid					●																	
Phosphoric acid (aerated)				●																		
Polythionic acid				●																		
Potassium carbonate			●																			
Potassium chloride				●	●			●	●	●												
Potassium chromate										●												
Potassium hydroxide			●	●	●								●									
Potassium permanganate										●									●			
Silver nitrate			●																			
Sodium aluminate			●																			
Sodium bisulfate				●	●																	
Sodium bisulfite				●	●																	
Sodium carbonate				●	●																	
Sodium chloride				●	●																●	
Sodium fluoride			●	●																		
Sodium hydroxide	●	●	●	●	●	●						●	●	●		●						
Sodium nitrate			●									●	●									
Sodium phosphate (tribasic)				●	●																	
Sodium sulfate				●																		
Sodium sulfide														●								
Steam														●				●				
Steam, geothermal (aerated)				●	●																	
Strontium nitrate			●																			
Sulfate liquor			●	●																		
Sulfonated oil													●	●								
Sulfur												●										
Sulfuric acid			●																			
Sulfuric acid + chlorine				●	●																	
Sulfurous acid					●																	
Toluene									●	●												
Triethanolamine										●												
Trimethylamine									●													
Uranyl sulfate																					●	
Water + hydrogen sulfide + light hydrocarbons					●																	
Zinc chloride				●	●																	
Zinc fluorosilicate													●									

FOOTNOTES:

*Key to Materials

Ferrous Alloys

1. Gray cast iron (1.1-2.8 Si)
2. Cast iron (14-32 Ni)
3. Low-carbon steel
4. AISI type 302; 304/304L (0.03 C max); 321; 347
5. AISI type 316/316L (0.03 C max); 317/317L (0.03 C max)
6. ACI Alloy 20 (19-20 Cr, 28-30 Ni, <0.75 Mn, >2 Mo, 4-4.5 Cu); CN-20 (19-20 Cr, 28-30 Ni, 3 Mo, 1.75 Cu)
7. AISI type 405; 410

Copper-Based Alloys

8. Copper No. 100-150; Copper Alloy No. 210; 220, 230; 502-546
9. Copper Alloy No. 240; 260
10. Copper Alloy No. 268; 270; 280; 464-467; 612; 687
11. Copper Alloy No. 706; 710; 715

Nickel-Based Alloys

12. 200/200L (99.5 Ni, 0.25 Mn, 0.15 Fe, 0.05 Si, 0.05 Cu, 0.06 C (alloy 200) or 0.02 C (alloy 200L)
13. Monel 400 (66 Ni, 31.5 Cu, 1.4 Fe)
14. Inconel 600 (76 Ni, 15.8 Cr, 7.2 Fe)
15. Incoloy 800
16. Hastelloy alloy B
17. Hastelloy alloy C

Metals

18. Aluminum
19. Gold
20. Lead
21. Titanium
22. Zirconium

Source: Corrosion Data Survey (Metals Section), 5th Ed.; National Assn. of Corrosion Engineers, Houston.

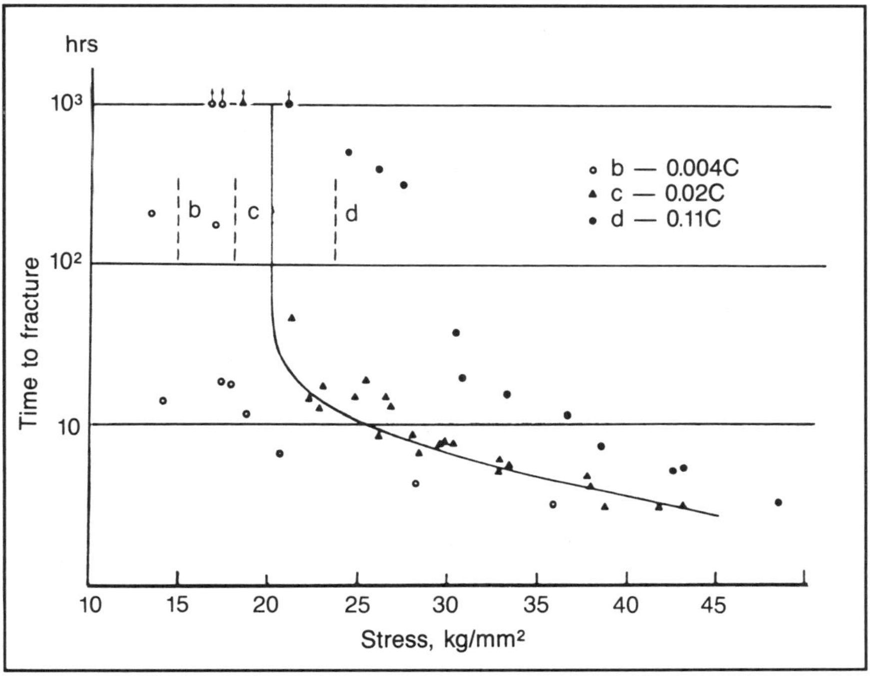

Figure 2–11 Stress dependence of several steels with different carbon content to stress-corrosion cracking in boiling 33% NaOH. Dotted lines b, c, and d represent the lower-yield strength of each steel (After Bohnenkamp).

every alloy is susceptible to cracking in every environment; rather, certain chemical species referred to as the specific ion are responsible for cracking each alloy. All metals and alloys are susceptible to SCC in some environment. Table 2–5 gives a description of some of these combinations but is by no means complete.

Changes in alloying content can significantly alter the cracking susceptibility of an alloy system. The classic example of this behavior is the systematic addition of nickel to iron-chromium alloys subjected to boiling 42% $MgCl_2$ (see Fig. 2–10). Here, the resistance goes through a minimum at approximately 8% Ni which corresponds to the 18-8 stainless steels. Microstructure is also of great importance. The presence or absence of precipitates and inclusions and the grain size strongly influence cracking resistance. Stress and temperature also influence cracking susceptibility. Increasing stress, temperature, and concentration of the specific ion act to decrease the stress-corrosion cracking life. Time to failure in a particular solution is logarithmically dependent on the applied stress in the same way discussed for fatigue. Figure 2–11 shows this relation for steel in an NaOH solution.

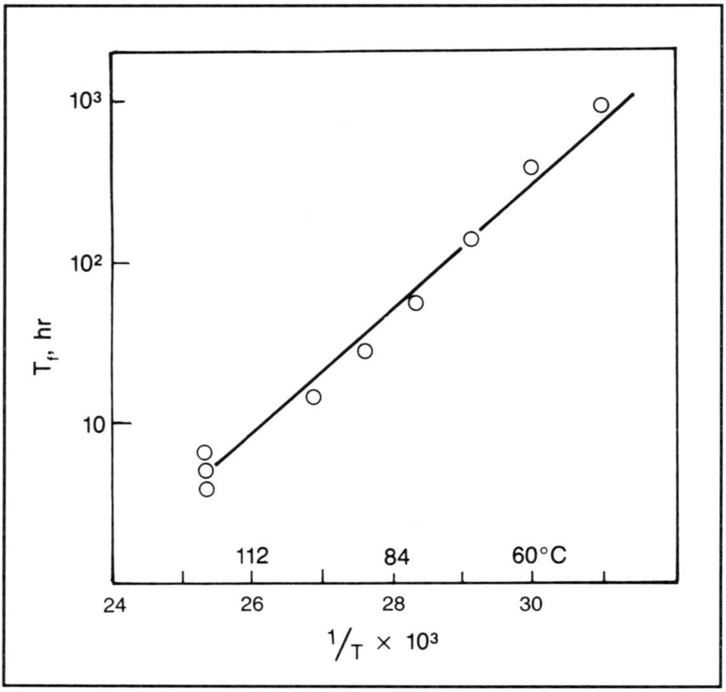

Figure 2–12 Effect of temperature on stress-corrosion cracking of steel in boiling 33% NaOH. Time to fracture (T_f) versus the reciprocal of absolute temperature is shown (After Bohnenkamp).

Similar to the fatigue case, there is often a threshold stress below which fracture will not occur. This same dependence is observed as a function of yield and tensile strength for high-strength steels. Increasing strength reduces the time to failure for stress corrosion cracking. Therefore, very high-strength steels can crack almost instantly when subjected to the right environment.

Increasing temperature also enhances cracking susceptibility (see Fig. 2–12). Note in this example that the changes in time to failure with temperature are logarithmic, so increasing the temperature from 72°C–112°C decreases the time to failure one order of magnitude. In many systems, there is a threshold temperature below which cracking does not occur as shown in Figure 2–13, which relates susceptibility to both concentration and temperature. At 100°F and below, no cracking occurs regardless of concentration of NaOH. Increasing concentration at constant temperature also increases cracking tendency as shown in Figure 2–13.

Similarities between stress-corrosion cracking and hydrogen damage have led

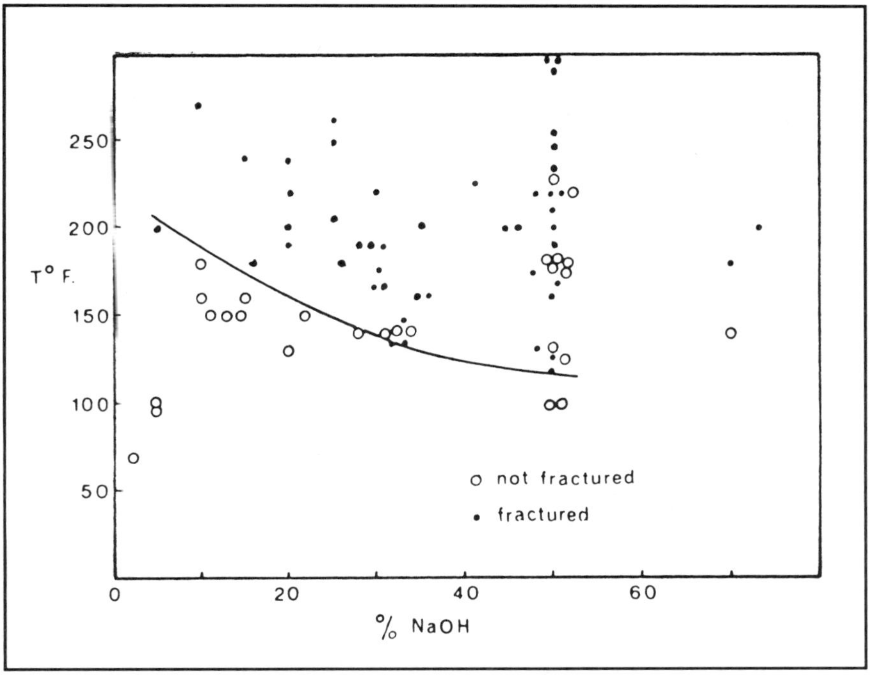

Figure 2-13 Effect of temperature and concentration on cracking of steel in NaOH solutions based on service experience (After Schmidt et al.).

some to believe that the two are related or that both participate in environmental cracking. Still other investigators adamantly contend that the two are separate, distinct mechanisms and cannot be related to a more general mechanism of environmentally assisted cracking.

One of the strong cases for the latter viewpoint is the use of applied potentials of either an anodic or cathodic nature during testing. It has been shown that application of a cathodic potential to a specimen cracking by an active path mechanism (stress-corrosion cracking) stops further cracking, while the application of an anodic potential accelerates cracking. The reverse was found for those specimens cracking presumably by hydrogen. However, recent work has shown that this method is not always effective and in fact produces rather anomalous results. Therefore, this method is not conclusive in distinguishing stress-corrosion cracking from hydrogen damage mechanisms but illustrates the complexity of these mechanisms.

Another method that has been used for determining which mechanism is operative is examination of the fracture path. A fracture path may be transgranular or intergranular in certain alloy systems in specific environments,

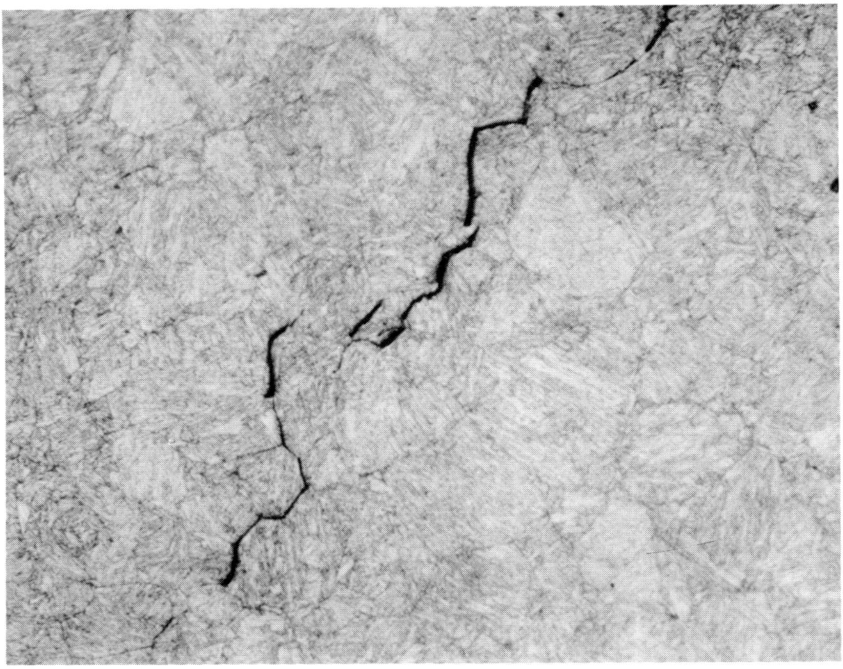

Figure 2–14 Intergranular fracture of Type 410 stainless steel bolt subjected to seawater environment. Magnified 165 times; etched with villela's reagent.

and from this, one mechanism can be differentiated from the other; however, more recent work has refuted this method since many factors such as alloy content, microstructure, heat treatment and stress affect the fracture path.

As an example of this and to aid the reader in understanding fracture path, Figures 2–14 – 2–17 are included. Figure 2–14 is representative of intergranular stress-corrosion cracking as it appears by optical microscopy, displaying a singular crack. Figure 2–15 shows typical branching-type stress corrosion cracking also by optical microscopy. The first example shows 410 stainless steel that cracked in a chloride environment (seawater) while the second shows 304 stainless steel that failed in a zinc bromide completion fluid. Figures 2–16 and 2–17 are included to acquaint the reader with scanning electron micrographs. The first shows an intergranular fracture of 4130 low-alloy steel from hydrogen stress cracking; the second figure shows the same alloy but with a transgranular fracture.

Since there is no consensus on the mechanisms of hydrogen damage and stress-corrosion cracking (SCC) or how to distinguish them readily, hydrogen damage is discussed as a separate environmental cracking mechanism. There is,

Figure 2–15 Transgranular fracture of a Type 304 stainless steel gravel pack screen as a result of exposure to $ZnBr_2$ completion fluid. Etched with oxalic acid; magnified 150 times.

in fact, good support for this distinction since several phenomena are quite different for the two.

There are many manifestations of hydrogen damage (HD), several of which are common to the oilfield such as sulfide-stress cracking (SSC), hydrogen-induced cracking (HIC), blistering, and in special cases, hydriding. More information on the mechanisms involved in these and other forms of hydrogen damage can be found in Volume 13 of the *Metals Handbook* referenced at the end of this chapter. One of the most common forms of HD in the oilfield is hydrogen-stress cracking (HSC) for which SSC is a special case. HSC is often initiated below the surface at discontinuities in the metal such as inclusions, grain boundaries, or locations below a notch that represent a region of high triaxial stress. However, SCC is always initiated at a surface notch or crevice which is commonly a pit. Thus, susceptibility to SCC is often related to a metal's susceptibility to pitting in an environment, while HSC is a function of the hydrogen absorption characteristics of the metal in the environment. Furthermore, the temperature dependence of cracking from HSC reaches a maximum around room temperature and is not a continually escalating function of temperature as is SCC. Figure 2–18 shows

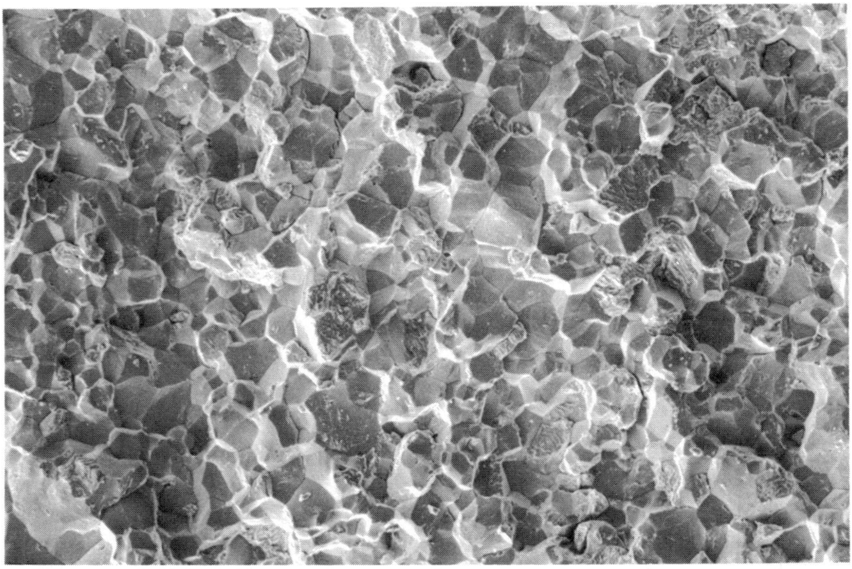

Figure 2–16 Scanning electron micrograph of fracture surface of AISI 4130 low-alloy steel. Failed in hydrogen, showing intergranular path of fracture magnified 200 times. Note crystal facets and rock candy appearance.

this behavior, which is in stark contrast to that shown in Figure 2–12 for SCC. This figure also demonstrates the effect of how rapidly the specimen is loaded. The minimum in resistance to hydrogen at approximately 25°C (room temperature) has important ramifications for petroleum production metallurgy (Chapter Four).

Some of the similarities between HSC and SCC are the effects of stress, concentration, alloy composition, and microstructure. The effect of applied stress is presented in Figure 2–19. This type of curve is commonly referred to as a static fatigue curve because of its similarity to the fatigue S-N curves; however, in this case the specimen is subjected to a static load. As in stress-corrosion cracking, the lower the applied stress, the longer the life to failure; below the threshold stress, failure does not occur.

HSC susceptibility is also a strong function of yield and tensile strength. In general, steels below 80,000 psi yield and 90,000 psi tensile strength are relatively resistant to cracking, although some cases of much lower strength failures have been reported. This threshold is also reported as Rockwell C22, Vickers 248, and Brinell 235. However, this is only a guideline value and not an absolute for design purposes.

As regards HSC, hydrogen ion concentration is the principal component of

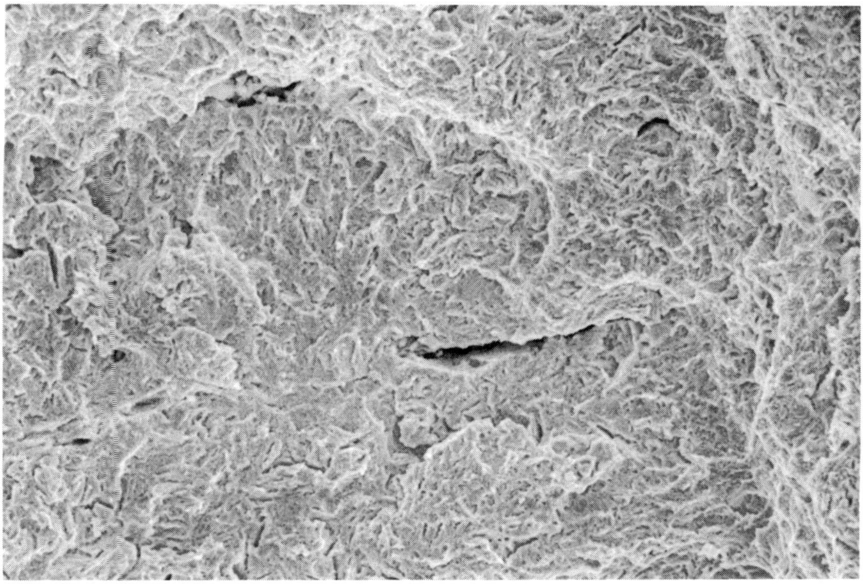

Figure 2–17 Same alloy as Figure 2–16 except showing transgranular fracture. Magnified 600 times.

concern in the environment and when reduced on the surface to hydrogen atoms is absorbed as hydrogen into the metal; thus, pH has a strong influence on HSC. Most corrosion reactions liberate hydrogen at the cathode so that hydrogen is free to either enter the metal or combine to form molecular hydrogen. Some environments such as those containing hydrogen sulfide enhance the ingress of hydrogen by poisoning the hydrogen recombination reaction to molecular hydrogen. The hydrogen molecule is not detrimental to alloys; only the atomic or nascent hydrogen is a problem. Figure 2–20 shows the effect of hydrogen content and tensile strength on the loss in ductility of steel.

Alloy composition is also an important consideration on the behavior of different alloys in the same hydrogen environment. Higher alloying usually imparts greater resistance to hydrogen except when accompanied with increasing strength. For instance, the austenitic stainless steels are typically more resistant than ferritic steels to HSC except when they are highly cold-worked.

Metallurgical structure also significantly affects the resistance of alloys to HSC. Typically, single-phase structures are more resistant to hydrogen, but there are exceptions such as untempered martensite. In steels at equivalent strength levels, tempered martensite is the most resistant to hydrogen. The bainitic and pearlitic structures have varying degrees of resistance depending on how fine the pearlite is and whether the bainite is upper or lower. Lower bainite

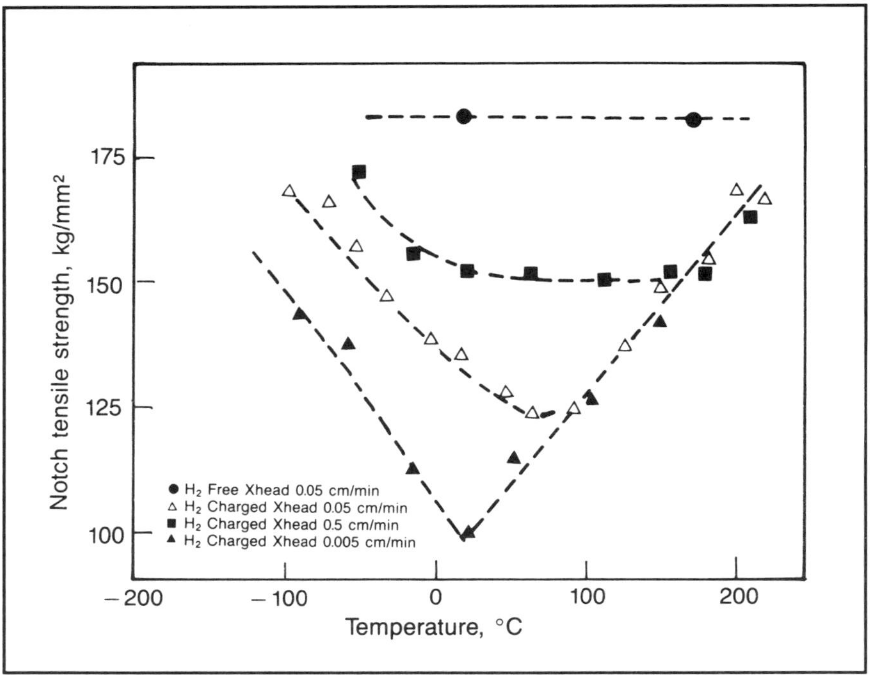

Figure 2–18 Reduction in tensile strength of hydrogen-charged steel as a function of temperature at three different loading rates (After Graville et al.).

is more resistant than upper bainite to HSC. In general, quenched and highly tempered steels are more resistant to HSC than ones that are normalized or normalized and tempered.

Hydrogen-stress cracking and stress-corrosion cracking have become increasingly important considerations with the advent and use of high-strength and high-alloy tubulars for production. These mechanisms will be discussed in greater detail in Chapter Four.

Methods of Prevention

The four most common methods of preventing or reducing corrosion in the oilfield are coatings, inhibitors, cathodic protection, and alloying (metallurgy). Since this book is about metallurgy in the oilfield, the first three topics will only briefly be discussed while the fourth will be emphasized in subsequent chapters.

Inorganic, organic or metallic coatings alter the corrosion process by chang-

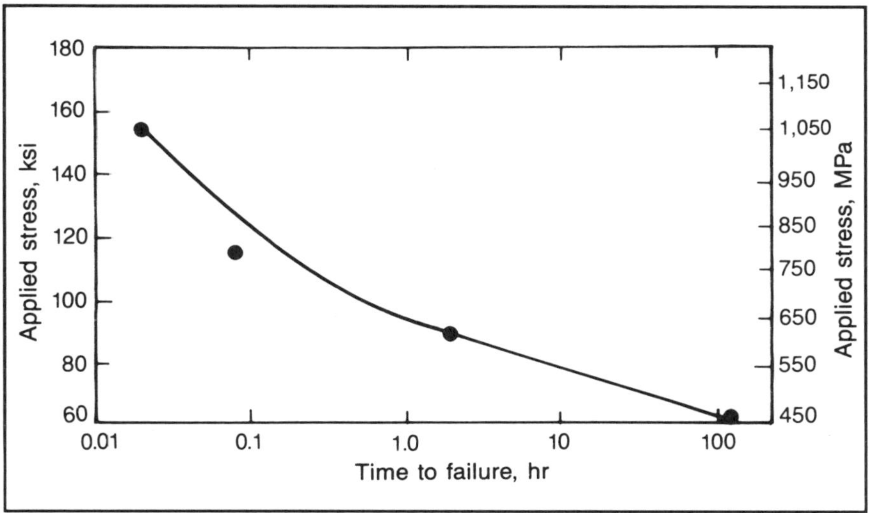

Figure 2–19 Effect of applied stress on hydrogen cracking resistance of 4130 steel.

ing the nature of the metal/solution interface. Organic and inorganic coatings accomplish this by providing a temporary barrier to the solution/metal reaction interface. It must be remembered that all these coatings are permeable to water, chlorides, hydrogen, and other gases such as carbon dioxide and hydrogen sulfide; thus, coatings represent a temporary although often sufficient solution to corrosion problems.

Metallic coatings are of two basic types: noble and sacrificial. An example of the first is nickel-plated steel. This plate offers a more corrosion-resistant surface in contact with many environments than the steel surface does. However, when the plate is locally removed or damaged by some process such as erosion, a strong galvanic effect may be initiated as a result of the very small anode (steel) and the large cathode (nickel) and may lead to rapid deterioration of the substrate. The second metal coating has already been explained by the example of galvanized steel earlier in this chapter.

Inhibitors represent another method of corrosion control. Inhibitors in this context include any chemical that reduces the tendency for corrosion. Therefore, oxygen scavengers, passivators, biocides, etc., are all included. This method of control is quite complex and discussion of even the fundamentals is not appropriate to this book; however, it represents one of the primary mechanisms for controlling corrosion in the oilfield. Basically, the use of chemicals can alter the corrosion environment by reacting with the corrodents, by forming a transient barrier much as coatings do, or by changing the local potential of the metal

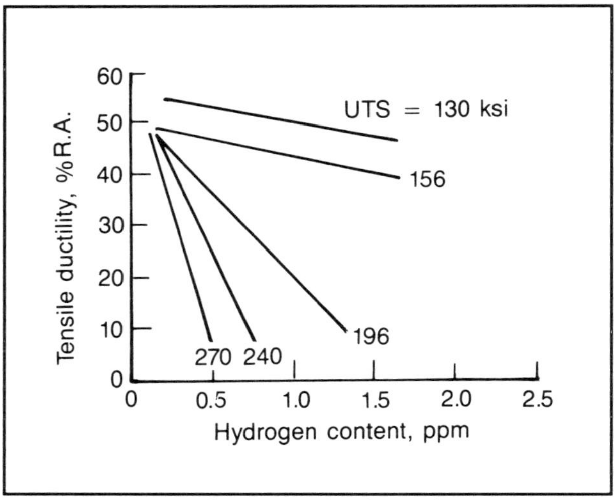

Figure 2–20 Reduction in ductility with increasing hydrogen content and strength (After Tetelman).

surface to a more noble potential, thereby reducing its tendency to corrode in a particular environment.

Cathodic protection involves changing the corrosion cell such that the component that needs protecting becomes the cathode at the expense of another metal, which becomes the anode. The simplest form of cathodic protection involves the direct contact between a metal that must be protected and one that is more active and can be sacrificed. Sacrificial anodes for protecting steel structures are usually alloys of zinc, aluminum, or magnesium. In most water and soil environments, these metals are anodic to steel and, when coupled, will preferentially corrode, protecting the steel. For environments in which longer protection times are required or more current must be applied than can be supplied by a sacrificial anode such as occurs in soil, an impressed current system may be used. This type of cathodic protection uses a rectifier and anodes that are usually more noble than the steel. The current is driven to the structure from the anode to cause the structure to become cathodic to the anodes.

Cathodic protection systems are widely used for external protection of pipelines, tanks, well casing, etc. Internal cathodic protection is frequently used in separators and storage tanks.

Materials selection in the form of alloy changes to provide resistance to corrosion is the thrust of this book and will be discussed in detail in remaining chapters. Historically, the drilling and production area of the petroleum industry has not easily adopted metallurgy as a solution to corrosion problems. But with the need to produce oil and gas from more aggressive environments, metallurgical solutions have become more common.

References

ASM International. *Metals Handbook,* Volume 10, 1975, p. 176.

ASM International, *Metals Handbook,* Volume 13, p. 1987, p. 163.

____. "Evaluating Media that Cause Stress-Corrosion Cracking." *Metal Progress Data Sheet,* Vol. 107 (1975), p. 59.

Betz Handbook of Industrial Water Conditioning, 1976, p. 183.

Bohnenkamp, K. "Caustic Cracking of Mild Steel." *Fundamental Aspects of Stress-Corrosion Cracking.* Houston, Texas: NACE, 1969, p. 374.

Copson, H.R. "Effects of Composition on Stress-Corrosion Cracking of Some Alloys Containing Nickel." *Physical Metallurgy of Stress-Corrosion Fracture.* New York, New York; Interscience Publishers Inc., 1959.

Craig, B.D., "Fundamental Aspects of Corrosion Films in Corrosion Science," Plenum Press, N.Y. 1991.

Fontana, M.G. and N.D. Greene. *Corrosion Engineering.* New York, New York; McGraw-Hill Book Company, 1967.

Graville, B.A.; R.G. Baker; and F. Watkinson. *British Welding Journal,* 14 (1967), p. 337.

National Association of Corrosion Engineers. *Basic Corrosion Course.* Eleventh Printing. 1980, pp. 2-7.

Schmidt, H.W.; P.J. Gegner, C. Heinemann, C.F. Pogacar, and E.H. Wyche. *Corrosion,* 7 (1951), p. 295.

Stolica, N.D. "Pitting Corrosion on Fe-Cr and Fe-Cr-Ni Alloys." *Corrosion Science,* 9 (1969), p. 455.

Tetelman, A.S. "Recent Developments in Classical (Internal) Hydrogen Embrittlement." *Hydrogen in Metals.* ASM, 1973, p. 17.

Uhlig, H.H. *Corrosion and Corrosion Control.* New York, New York: John Wiley and Sons Inc., 1967.

3

METALLURGY IN DRILLING

A multitude of special tools and equipment used in drilling oil and gas wells utilize many different alloys. The metallurgy of this equipment has evolved over many years and is often proprietary and therefore not readily available from the literature or the manufacturer. This chapter will present information on the more common equipment used in rotary drilling such as drillpipe, tool joints, drill collars, and blowout preventers (BOP).

Even limiting the discussion to these components presents a broad array of manufacturers' preferences for materials. Therefore, the examples presented do not provide a complete analysis of all of the alloys being used for a specific component; rather, they represent the most common alloy selection and some of the reasons for these choices.

Metallurgy of Drilling Equipment

The emphasis in drilling equipment is on pressure containment and fracture resistance—the latter concerning low-temperature fracture of surface equipment, fatigue of subsurface components, and sulfide-stress cracking (SSC) for both surface and subsurface equipment. Fine grain size and a quenched and tempered martensitic microstructure offer the greatest resistance to fracture and the highest strength for pressure containment. Thus, most equipment metallurgy is selected with these features in mind. This, of course, requires adequate hardenability for components that are constructed of heavy sections. Some of these are on the order of 6 in. or greater in wall thickness and several tons of mass. Therefore, it is quite common to use 4100 and 4300 series alloys because of their good hardenability, but to keep the carbon content as low as possible from a weldability standpoint since some weld repair is likely on most cast components.

Pressure control at the surface during drilling or workovers is attained using blowout preventers (BOP). A typical BOP stack (see Fig. 3–1) consists of a

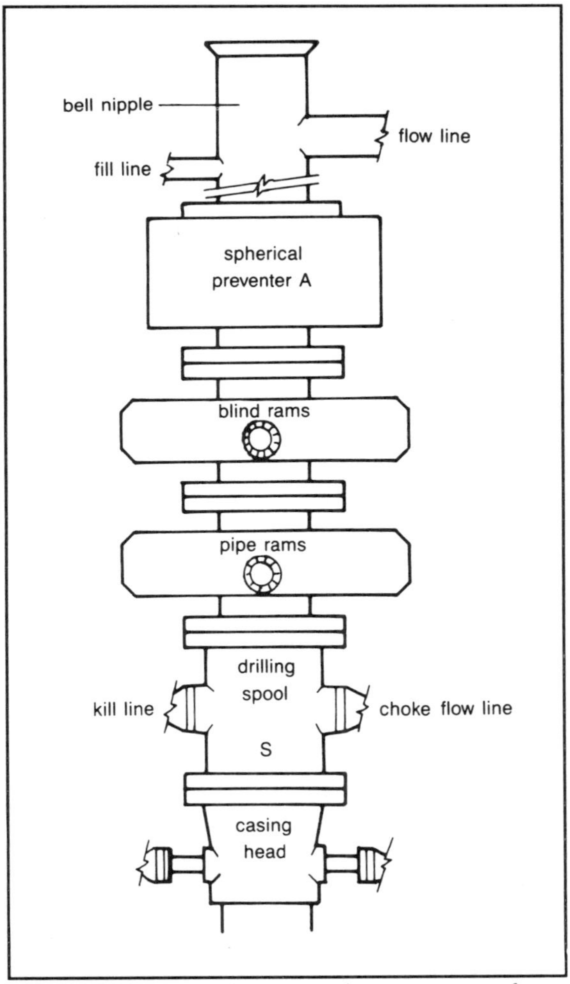

Figure 3–1 Typical BOP stack arrangement for drilling.

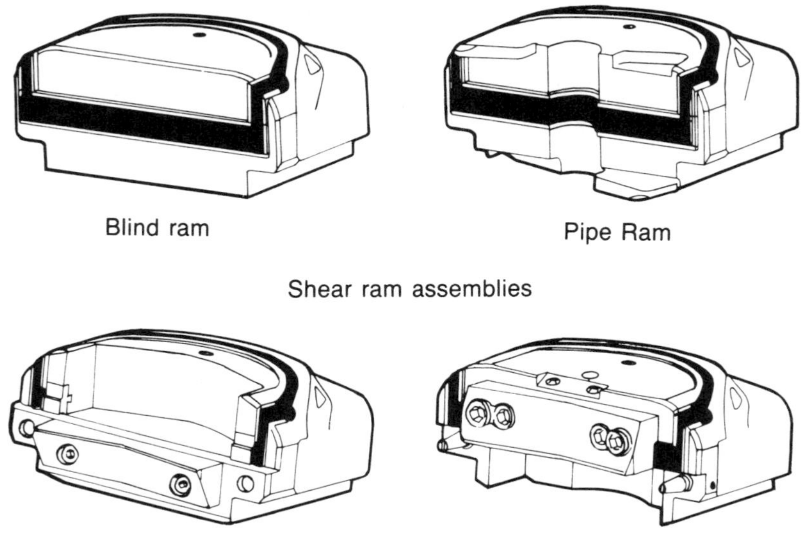

Figure 3–2 Blind, pipe, and shear rams used in ram-type BOP (courtesy Hydril Company).

spherical or annular preventer, ram preventers, a drilling spool, and a casinghead.

The spherical preventer seals around the outside of the kelly, tool joint, or drillpipe in the bore or can be closed on an open hole. Sealing is accomplished using elastomeric material. The operating cylinder, which moves the sealing material, typically is made of quenched and tempered 4130 steel. The body of the preventer is usually cast or forged low-alloy steel such as AISI 4130 or 4140.

There are three basic types of ram preventers: shear, blind, and pipe. Since ram bodies are universal, only the rams are changed to comply with the size pipe being run. These bodies are most often made of low-alloy cast or forged steel of the 4130 type. The different types of rams are shown in Figure 3–2. Shear rams are used to shear the pipe completely, sealing off the annulus and the drillpipe. To achieve this shearing quickly and easily, which may entail shearing the high-strength tool joint (140,000 psi minimum tensile strength), the shear blades are made of 4340, 4140, or 2 1/4 Cr-1 Mo steel quenched and tempered to a hardness of Rockwell C 45. Blind rams are used to seal off an open hole with no pipe in it, and pipe rams seal around pipe in the hole to isolate the annulus. Since shearing is not necessary on these rams, the hardness levels can be

Table 3–1 API Material Application Pressure Containing Members

Part	Pressure Ratings (psi) 2000	3000	5000	10,000	15,000	20,000
	API Material Designation					
Body	36K.45K. 60K. 75K.	36K. 45K. 60K. 75K.	36K. 45K. 60K. 75K.	36K. 45K. 60K. 75K.	45K. 60K. 75K.	60K. 75K
Integral End Connecton	60K	60K	60K	60K	75K	75K
Blind Flange	60K	60K	60K	60K	75K	75K
Blind Hub	60K	60K	60K	60K	75K	75K

(courtesy API)

reduced but still may exceed the National Association of Corrosion Engineers' (NACE) requirements for resistance to sulfide-stress cracking. Pistons that drive the rams are usually made of quenched and tempered 4140 or 4340 steel.

Materials for different components of these BOPs are specified by the API in API Specification 16A, *"Drill Through Equipment,"* and *Specification 6A: Wellhead Equipment.* Table 3–1 shows that the type of material used is dependent on the working pressure of the equipment. The different types of material are shown in Table 3–2. The carbon and low-alloy steel compositions cover common alloys of construction such as 4130, 4140, 8620, 8630, and 2 1/4 Cr-1 Mo. The martensitic stainless steels cover AISI 410, F6NM and CA6NM.

Table 3–2 Steel Composition Limits (WT%) Material for Pressure Containing Members

Alloying Elements	Carbon and Low Alloy Steels Composition Limits	Martensitic Stainless Steels Composition Limits
Carbon	0.45 Max	0.15 Max
Manganese	1.80 Max	1.00 Max
Silicon	1.00 Max	1.50 Max
Phosphorus	0.04 Max	0.04 Max
Sulfur	0.04 Max	0.04 Max
Nickel	1.00 Max	4.50 Max
Chromium	2.75 Max	11.0–14.0
Molybdenum	1.50 Max	1.00 MAx
Vanadium	0.30 Max	NA

(courtesy API)

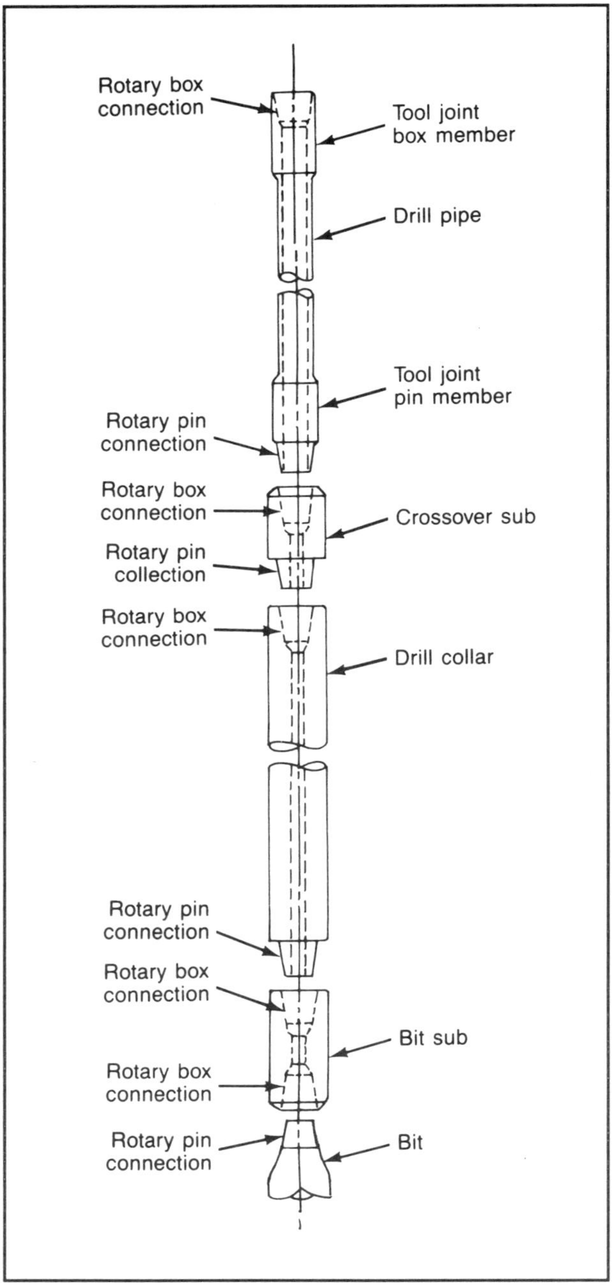

Figure 3–3 Typical drillstem assembly (from *API Specification 7*).

Bolting on preventers is covered by API *Recommended Practice 53* and, depending on whether the flange type is 6B or 6BX, the nuts are made of ASTM A 194 Grade 1 or 2H, respectively. In both cases, the bolts are made of either ASTM A 193 Grade B and/or A 354 Grade BC.

All material for BOPs must have a minimum average Charpy impact value of 15.0 ft-lb for a set of three full-sized specimens with a minimum of 10.0 ft-lb for any one specimen of the set. It is required for the user to specify a temperature, -0°F, -20°F, or -75°F, at which these values are required to be met.

In cases where an environment containing hydrogen sulfide is anticipated, the maximum hardness for the preventer bodies is HRC 22, and all flange and bonnet bolts should be made of ASTM A 193 Grade B-7M with a hardness of HRC 22 or less and nuts to ASTM A 194 Grade 2M, according to API RP53. API 6A also allows Monel K500 studs in the hot-rolled and age-hardened condition to a maximum hardness of HRC 35 and ASTM A 453 Grade 660 (A 286) solution-treated and age-hardened to HRC 35 maximum for both bolts and nuts.

Flanges can be made from ASTM A 105 forgings, A 216-WCB castings, or API Type 2 or Type 4 material.

The drilling spool and casinghead are often made from line pipe specified in API 5L, or from casing specified in API 5CT. The mechanical properties of these items are matched to those of the preventers and casing. Sometimes casingheads are constructed of 4130 or ASTM A 487-9Q steel and are quenched and tempered when special environments such as hydrogen sulfide or low temperature require specific properties. For sulfide-containing environments, controlled hardness to HRC 22 maximum is required. In the case of low-temperature service, notch toughness is enhanced by quenching and tempering.

The *drillstem,* which is shown in Figure 3–3, is probably the most studied assembly in drilling. Tool joints, drill collars, subs, and drillpipe form the basis of the remaining discussion in this chapter. Drill bits are not discussed because they represent a very specialized component for which little choice in material selection by the user is available.

Tool joints are the end connections of drillpipe and are made by forging. The final heat-treated joint, which is quenched and tempered, is friction-welded, inertia-welded or flash-welded to the pipe body. In friction welding and inertia welding, the pipe is held stationary while the tool joint is rotated onto the pipe body; the resulting contact pressure causes temperature elevation with subsequent welding. For flash welding, both the pipe and the tool joint remain stationary while an electrical current is impressed across the joint. Heat is generated by electric resistance at the interface and welding results. After both processes, postweld heat treatment is applied to the weld zone to improve uniformity of structure and properties. This heat treatment can take the form of a simple high-temperature stress relief, or by induction-heating the weld-heat-affected zone, liquid quench (usually with water), then induction-temper. Historically, these welds have been quite sound with relatively few failures. However, failures of the tool-joint weld have been observed.

Tool-joint material is commonly AISI 4100-series low-alloy steel, usually 4135-4140, although 4145H is also used. For heavy-wall drill pipe, AISI 1340 tool joints are often used. Recently, 4100-series steels with molybdenum and niobium modifications have been introduced because of their greater resistance to cracking in hydrogen sulfide environments. Tool joints are normally quenched and tempered to a hardness of 30-37 Rockwell C, resulting in yield strengths of 120-150 ksi. Tempering is performed for 2 hr at 1,100-1,200°F. API *Specification 7,* section 4, requires a minimum yield strength of 120,000 psi, a minimum tensile strength of 140,000 psi, and a minimum elongation of 13% for all new tool joints of all sizes. The yield strength is measured by the 0.2% offset method in comparison to the 0.5% extension method used for tubulars. Small joints that do not provide for easy tensile-specimen removal can be Brinell tested. A minimum Brinell hardness number (BHN) of 285 is acceptable on both the box and the pin.

Drill collars are also made of 4135-4140 or 4145H steel quenched and tempered to a hardness range of 285-341 BHN. Drill collars are used to add weight to the string and to keep it in tension so the effects of fatigue are minimized. Minimum tensile requirements for drill collars in API *Specification 7* are for OD ranges of 3 ⅛–6 ⅞ in. The yield is 110,000 psi, tensile strength is 140,000 psi and the elongation is 13%; for 7–10 in. OD collars, the yield is 100,000 psi, tensile strength is 135,000 psi, and minimum elongation is 13%. Similar to tool joints, the yield is measured by the 0.2% offset method.

Nonmagnetic drill collars are placed in a string during directional drilling to help determine the actual location of the string and bit. These collars are made from a variety of alloys, the most common of which are Monel K500, 316L austenitic stainless steel, several other austenitic stainless steels and beryllium copper. Austenitic stainless steel collars have failed in drilling environments from chloride stress-corrosion cracking reportedly by high residual stresses from manufacturing or sensitization of the collar. Care should be taken when using austenitic stainless steel collars in hot brine drilling fluids.

Drilling jars, stabilizers, subs and, quite often, core barrels are made from 4140 or 4145H steel and heat-treated as described for drill collars and tool joints. Sometimes AISI 4340 or 4340H steel is used for the metallurgy of these items and it too is quenched and tempered, typically to the same or higher hardness level.

Drillpipe is the last part of the drillstem mentioned here but is far from the least. In terms of research and manufacturing, this component represents the most-studied component and often the area of greatest trouble in drilling.

The various grades of drillpipe, according to the API, are shown in Table 3–3. Steels used for drillpipe are of the C–Mn or Cr–Mo type, the latter being 4100-series steel. Also, to a lesser degree, aluminum drillpipe is used, which is typically Type 2014, an aluminum-copper alloy. The steels are provided in the normalized and tempered or quenched and tempered condition, while the aluminum pipe is heat-treated to the T6 condition.

Table 3–3 API Requirements for Drillpipe*

	Yield Strength, ksi (MPa)		Tensile Strength, ksi (MPa)	Chemistry %		
Grade	min.	max.	min.	P (max.)	S (max.)	Heat Treatment**
E	75 (517)	105 (724)	100 (689)	0.030	0.030	Heat Treated
X–95	95 (655)	125 (862)	105 (724)	0.030	0.030	Q&T/N&T
G–105	105 (724)	135 (931)	115 (793)	0.030	0.030	Q&T/N&T
S–135	135 (931)	165 (1,138)	145 (1,000)	0.030	0.030	Q&T/N&T

* Elongation requirements depend on pipe thickness and therefore are not shown
** Q&T — quenched and tempered
N&T — normalized and tempered
(From API *Specification* 5D)

Under the requirements of API *Specification 5D,* drillpipe can only be seamless pipe and must be heat-treated full length, or re-heat-treated if upset. Heat treatment for high-strength series drillpipe is specified as either quenched and tempered or normalized and tempered. The only chemical composition requirements are that the phosphorus and sulfur contents not exceed 0.030 and 0.030% each.

Grade E drillpipe is normalized or normalized and tempered and has a chemistry similar to 1041 or 1045 steel modified with about 1.5%Mn and 0.2% Mo. Grades G, and S are usually quenched and tempered. Grade X is typically normalized and tempered and has a chemistry range of 0.20–0.30%C, 1.20–1.50%Mn, 0.4–0.6%Cr, and 0.2–0.5% Mo. Grade G has a chemistry similar to X, while S-135 is typically 0.27–0.35%C, 1.50–1.60%Mn, 0.1–0.5%Cr, 0.3–0.4% Mo, and 0.012–0.016%V. Heavy-wall drillpipe is made from 4140 and 4145H quenched and tempered steel. Rockwell C hardness of the various grades ranges from 20–28 for Grade E, 27–30 for X-95, 30–34 for Grade G, and 34–37 for S-135.

Aluminum drillpipe as mentioned is frequently made of Type 2014 aluminum, which has the composition 0.5–1.2%Si, 1.% Fe max, 3.9–5.0%Cu, 0.4–1.2%Mn, 0.2–0.8%Mg, 0.10%Cr max, 0.25%Zn max, and 0.05%Ti max. Typical mechanical properties when heat-treated to the T6 condition are 64 ksi tensile strength, 58 ksi yield strength, 7% elongation, and a BHN of 135. Aluminum drillpipe usually comes with steel tool joints that are threaded on to ensure maximum joint strength that cannot be attained with aluminum tool joints.

Causes and Prevention of Drilling Equipment Failures

Reports in the literature of BOP failures are rare, probably for two reasons. First, since these units represent the primary method of preventing potential blowouts with an uncontrolled kick in a well, they must be over designed and of good structural integrity. Second, fewer failures occur on surface equipment compared with the drillstem, which sees more severe environments in terms of corrosion and complexity of stress state. However, sulfide-stress cracking (SSC) and brittle fracture of blowout preventer bolts has been reported in the past. For the most part, these were documented as being out of specification as far as API and NACE requirements.

Tool joints and drillpipe represent the area of greatest concentration over the past several decades, and thus, more is known and documented about their behavior. The first in-depth paper on this subject, "Causes and Prevention of Drill-Pipe and Tool-Joint Troubles," was written by R.S. Grant and H.G. Texter. This paper was last revised in 1948. No other comprehensive work on this subject has since been published, although other papers dealing with specific types of failures have been written. To introduce this multifaceted subject, purely mechanical failures will be dealt with before the actual drilling environment.

The stresses on a drillstring that is rotating during drilling—often in a deviated hole—are basically fourfold, three of which are shown schematically in Figure 3–4. There is a *tension load* caused by hook loads (weight of the pipe) and from the drill collars at the bottom of the hole. Depending on the number of drill collars used and the weight of the drilling mud, which adds buoyancy to the string, part of the drillstring will be in compression. The point at which the string changes from tension to compression is called the *neutral point*. Since drilling in compression decreases the life of the string faster than drilling in tension, the drill-collar weight is designed to move the neutral point as low in the hole as possible, allowing the largest portion of the string to be in tension. Bending stresses caused by drilling in a deviated hole cause cyclic compression and tension as the pipe rotates, alternating these two stress states. Finally, there is a torsional stress caused by rotation of the pipe when the bit is cutting

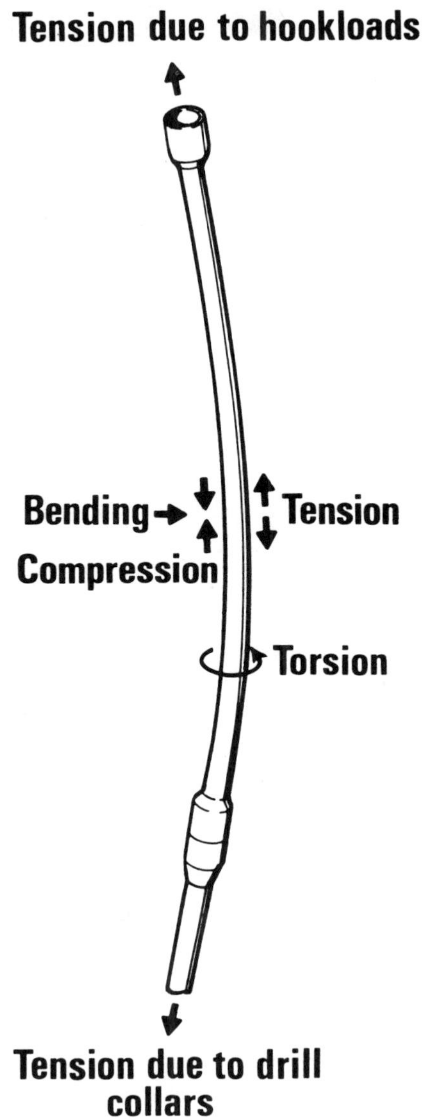

Figure 3–4 Three major stress conditions on drillstem: bending, torsion, and tension.

Figure 3–5 Drillpipe failure from tensile overload (after Grant and Texter).

into the formation. The fourth stress, not shown in Figure 3–4, is caused by vibration of the drillstem. This stress is difficult to quantitate and extremely complex. Vibration analysis is beyond the scope of this book, however, it is thought to be the cause of very few failures.

Twistoff is the mechanical failure of the pipe in torsion caused simply by overstressing the pipe; however, this term is often misused for any drillpipe failure that results in complete fracture of the pipe. Twistoff is more commonly used to describe fatigue failures. Torsional failure is rare but will occur if the bit becomes stuck and the rotary table continues to turn.

A more common failure than twistoff is *tensile failure.* This failure occurs when the pipe is pulled in two because of excessive stress on it. This happens most often when pipe is stuck in the hole and the drilling crew tries to pull on the pipe without monitoring the weight indicator. Pure tension failures are easily distinguished by the necking down (reduction) of the wall at the point of failure. The fracture will be circumferential and will produce a 45° angle along most of the surface. The OD of the pipe near the fracture will also have been reduced substantially. These features are shown in Figure 3–5.

Fatigue is the most common cause of failure in drillpipe and is variously called *washout* or *twistoff* in the field, depending on the appearance of the failure. A fatigue crack that has initiated in the pipe in the transverse direction, allowing drilling fluids to penetrate the wall and cutting out a hole, is called a washout. Figure 3–6 shows an example of a washout in drillpipe looking at the inside surface. If the fatigue crack propagates rapidly and the pipe parts, the failure is referred to as twistoff (see Fig. 3–7). Both of these terms are misno-

Figure 3–6 View of a washout (arrow) near the upset runout on the inside of the drillpipe.

mers but remain part of the oilfield jargon but, in fact, are both the result of fatigue. Recently, it has been demonstrated that one of the primary reasons why drillpipe either parts or washes out is due to the fracture toughness. Higher fracture toughness measured by Charpy impact produces washout whereas lower toughness causes twistoff. Figure 3–8 shows Charpy impact data of field samples for Grades E and X and also tool joints. It was determined from this work that a minimum Charpy impact of 40 ft-lbs at 75°F was sufficient to prevent twistoffs. The advantage of a washout over a twistoff is the shorter time required for tripping a string rather than fishing for a twistoff.

There are three types of fatigue: pure fatigue, notch fatigue, and corrosion fatigue. Often, these types are not distinct since corrosion fatigue is initiated at pits that provide a notch for fatigue crack initiation. Furthermore, it is now well

Figure 3–7 Corrosion-fatigue failure (twistoff) of drillpipe caused by internal pitting.

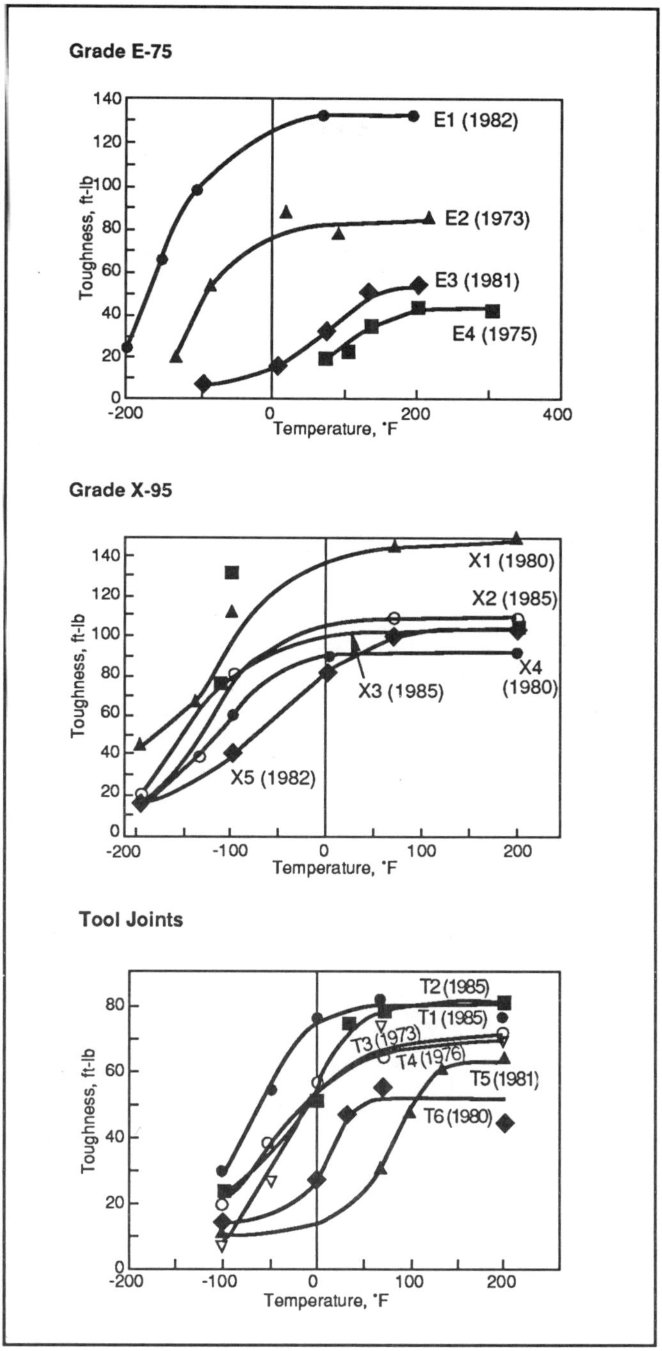

Figure 3–8 Charpy impact behavior of Grades E and X drillpipe and tool joints (after Buscemi et al.).

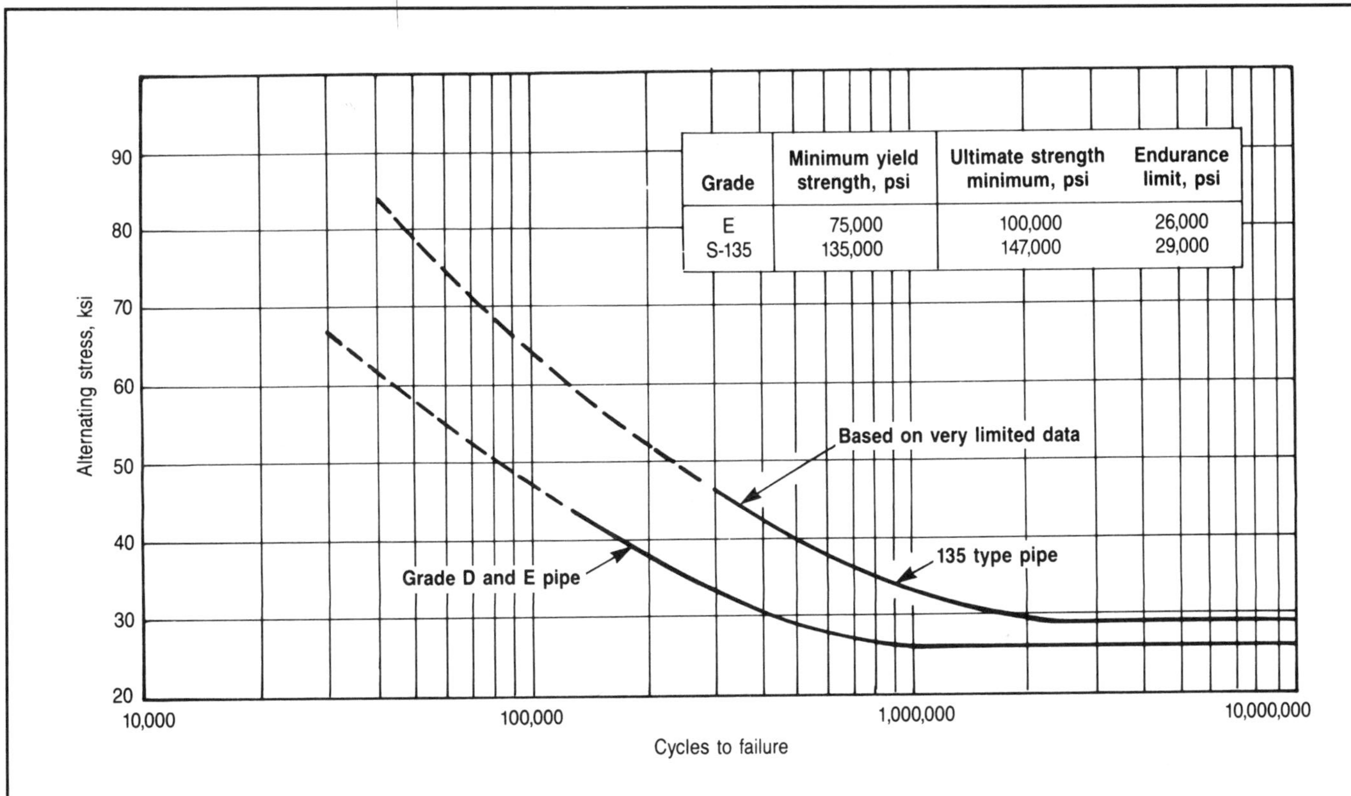

Figure 3–9 S-N curves for Grade E and Grade S-135 drillpipe; obtained by cyclic-bend full-scale test samples (after Mehdizadeh).

established that pure fatigue begins by micronotch formation within the metal. However, for discussion's sake each type will be described.

Pure fatigue was discussed in Chapter One and does not occur to any extent in the oilfield. What should be kept in mind is that fatigue damage is cumulative, and as such what appears to be good pipe after several holes may be close to crack initiation and failure. In fact, as Grant and Texter mention, drillpipe rotating at 100 rpm for just 10 days will have more than 1 million cycles imposed on it. In deep wells drilling time can run from months to more than 1 yr, imposing millions of stress cycles on a drillstring.

As discussed in Chapter One, the fatigue limit or endurance limit is, as a rule of thumb, ½ the tensile strength. Thus, it is commonly held that increasing the strength of a component will increase its fatigue strength. Figure 3–9 shows this to be true, but it is not practically significant. The tensile strength is increased almost 50%, while the endurance limit is increased only about 10%.

Fatigue strength is reduced in the presence of a notch, as shown in Figure 3–10. Notches act to raise the stress locally, often beyond the yield strength of the material. The effect of a notch is a function of its sharpness and orientation to the applied stress. Sharper notches increase the stress ahead of the notch root, making crack initiation easier than does a blunt notch. The depth of the notch is also a factor in fatigue life because deeper notches significantly reduce fatigue life in comparison to a shallow score mark. Slips or tongs are the most frequent causes of notches in drillpipe and often lead to fatigue failure, which initiates from the external surface of the pipe at the root of the notch.

Similar to notches, a rapid change in cross section from a heavy cross section to a thin cross section can lead to reduced fatigue life. Numerous drillpipe failures plagued the worldwide drilling industry in the mid-1980s with no obvious explanation. The fatigue failures were originating in the transition area of the internal upset where the drillpipe makes a transition from the tool joint upset (heavy section) to the drillpipe body (thin section). Subsequent research established that a short internal taper length, M_{iu}, decreased the fatigue life whereas a long generous taper increased the fatigue life. While finite element modeling and fatigue testing are required to determine the optimum length of M_{iu} for each size of drillpipe, Figure 3–11 shows the benefit of adequate internal upset length on the fatigue life of 5 inch IEU, 19.50 pound per foot Grade E drillpipe.

Corrosion fatigue is the leading cause of drillpipe failure. This kind of failure is the result of the combined effects of a notch formed by a corrosion pit, which acts as a stress concentration, and fatigue, that is also assisted by the corrosive environment. As shown in Figure 2–8 in a corrosive environment the S-N curve has a distinct slope rather than an endurance limit, and failure will ultimately occur when the requisite number of cycles are attained. As in the case of notch fatigue, the sharper the notch (pit) bottom, the more severe the stress concen-

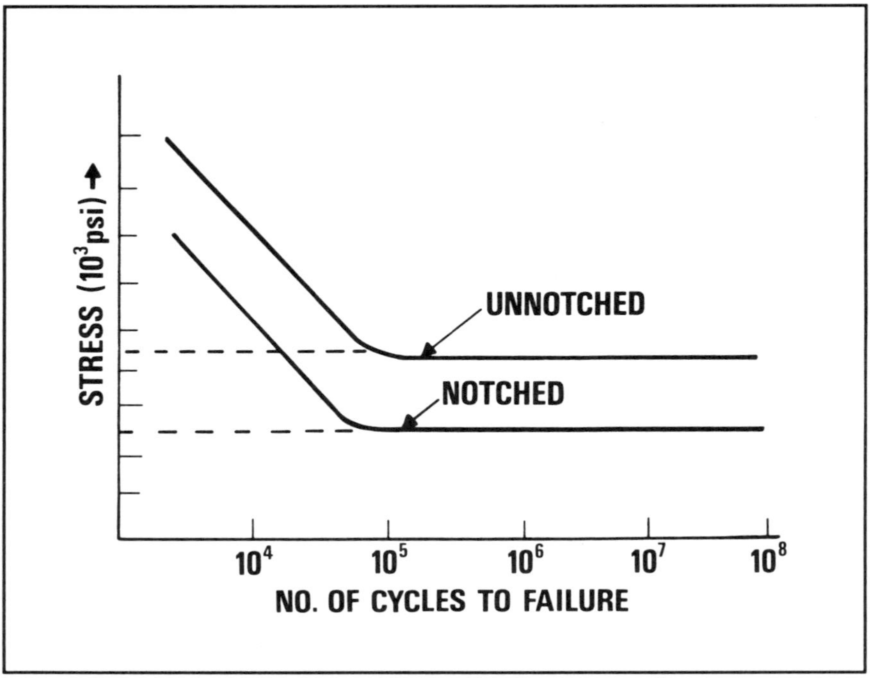

Figure 3–10 Effect of notches on fatigue resistance in steels.

tration and the shorter the fatigue life. This emphasizes the concept that corrosion fatigue is most often encountered in pitting environments. Since dissolved oxygen, chlorides, hydrogen sulfide, and carbon dioxide all promote pitting, it is easy to understand why drillpipe fails by this mechanism. As Grant and Texter showed for drillpipe samples tested in bending fatigue, the number of cycles to failure diminished rapidly with the increasing severity of the environment (see Table 3–4). Another view of this effect of environment is presented in Table 3–5

Table 3–4 Corrosion Fatigue of Drillpipe

Average Maximum Stress (psi)	Medium	Cycles to Failure (10^6)
31,500	Noncorrosive (air or distilled water)	2.20
31,500	Mildly corrosive (saltwater)	1.43
31,500	Corrosive (magnesium chloride)	0.86
31,500	Very corrosive (dilute HCl acid)	0.18

(After Grant and Texter)

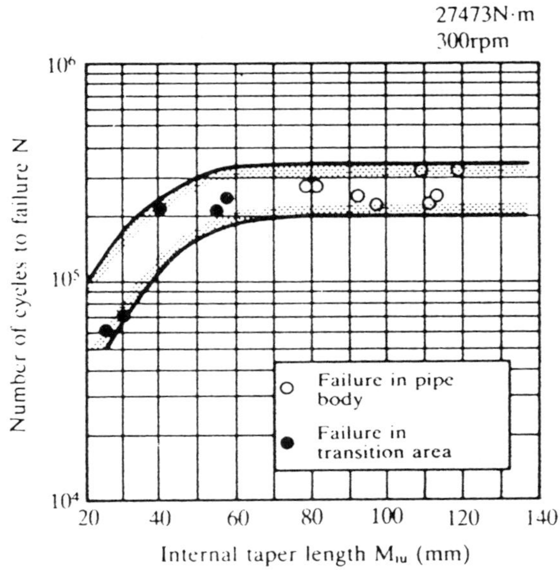

Figure 3–11 Effect of internal taper length on fatigue life of Grade E drillpipe (after Tsukano et al.).

from work by Mehdizadeh in which the affect of dissolved gases in saltwater on fatigue is described.

As discussed earlier, an example of corrosion fatigue is shown in Figure 3–7. The long, flat surface at the top of the pipe is the site where fatigue cracking occurred. The cracking was initiated by a large pit on the inside surface at the

Table 3-5 Corrosion Fatigue of Drillpipe

Dissolved Gas in Saltwater	% Decrease from Air Endurance Limit
H_2S	20
CO_2	41
CO_2 + Air	41
H_2S + Air	48
$H_2S + CO_2$	62
Aerated saltwater	65

(After Mehdizadeh et al.)

center of the fatigue-cracked region. After enough cross-sectional area of the pipe had been cracked, the remaining portion was not sufficient to support the tensile load, so final fracture of the remaining pipe wall resulted from simple tensile overload. A similar kind of fatigue fracture can occur from the outside if corrosion or erosion from cuttings in the mud allow pitting to progress to the extent necessary for fatigue crack initiation.

The greatest advantage of the drilling process over petroleum production is the ability to alter the environment (i.e., the drilling mud) and reduce potential problems created by the environment. By the use of chemical inhibitors and scavengers, corrosion can be reduced substantially, increasing the fatigue life of the drillstring. Quite often, simple pH control can extend the service life of a drillstring. This enhancement by pH control is shown in Figure 3–12. Increasing pH from 6.6, which is characteristic of most oilfield waters, to 12.1 and 13.0 restores the threshold or endurance limit in an aerated saltwater solution. By removing dissolved oxygen in drilling muds, pHs of 10–11 are normally sufficient to regain the greater portion of fatigue life.

At present, there are essentially no practical metallurgical solutions to corrosion fatigue. Most remedies are plastic coating the drillpipe and/or chemically treating the mud to reduce corrosion. Increasing the strength of the pipe does not significantly increase the fatigue strength, as shown in Figure 3–9, and in fact can lead to other problems when associated with hydrogen sulfide. Aluminum and titanium drillpipe offer metallurgical alternatives and will be discussed later.

In the presence of hydrogen sulfide, high-strength drillpipe can fail by a hydrogen-stress cracking mechanism referred to as sulfide-stress cracking (SSC). This form of failure is caused by the absorption of hydrogen into the metal from the corrosion reaction of steel and wet hydrogen sulfide. While the actual mechanism of cracking is in much dispute, the parameters that govern susceptibility to SSC are well documented. These factors are strength, applied stress, pH, H_2S concentration, temperature, and microstructure.

Increasing strength and applied stress increase SSC susceptibility as do decreasing temperature and pH. The best microstructure for resistance to SSC at comparable strength levels is highly tempered martensite, while the least-resistant is a normalized structure.

As expected, S-135 pipe is quite susceptible to SSC, while X-95 pipe can be used when the drilling mud is treated to remove hydrogen sulfide. In high concentrations of hydrogen sulfide in which scavenging may not be completely effective, the lower grades, E and X, are best. Since SSC resistance of steels reaches a minimum at or near 25°C, the greatest problem with cracking occurs at the top of the hole where high-strength pipe is most often run. To reduce this problem, tapered strings can be used to reduce the load on the top of the string,

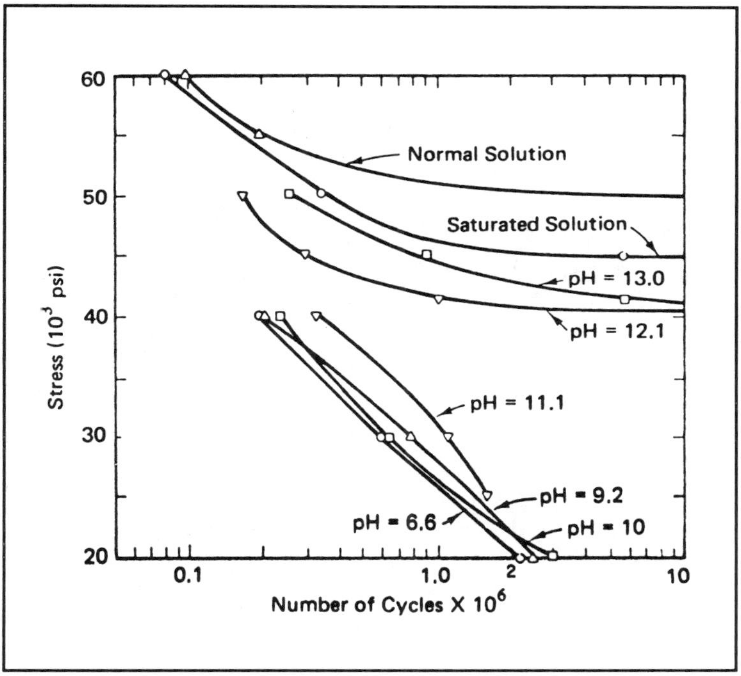

Figure 3–12 Effect of pH on corrosion fatigue in aerated saltwater (after McGlasson).

and thicker wall pipe can be run at the top to reduce the applied stress. Since X-95 steel has a hardness of 27–30 HRC and is often quenched and tempered, it usually stands a much better chance of resisting SSC than the higher-strength G-105 and S-135 steels. Figure 3–13 shows a typical SSC failure of S-135 pipe with the characteristic brittle fracture and the final ductile-shear lip. One very important consideration that is frequently ignored or forgotten is the high hardness of the tool joints. All the tool joints are of the same hardness which is essentially equivalent to S-135 drillpipe. Therefore, using lower strength drillpipe only reduces the potential of cracking of the pipe body not the tool joints. It is quite common in low strength drillpipe (E and X) for the tool joint pins to fail from SSC because they are the most highly stressed. Steel tool joints in general are particularly susceptible to SSC because of their higher strength. Most often pin failures occur where a fracture follows the circumference of the joint around the root of the last engaged thread. This region represents both the highest-applied stress and a stress concentration caused by the thread itself. Stress calculations and measurements have shown that the stress at the last engaged thread can achieve or exceed the yield strength. Although the boxes fail

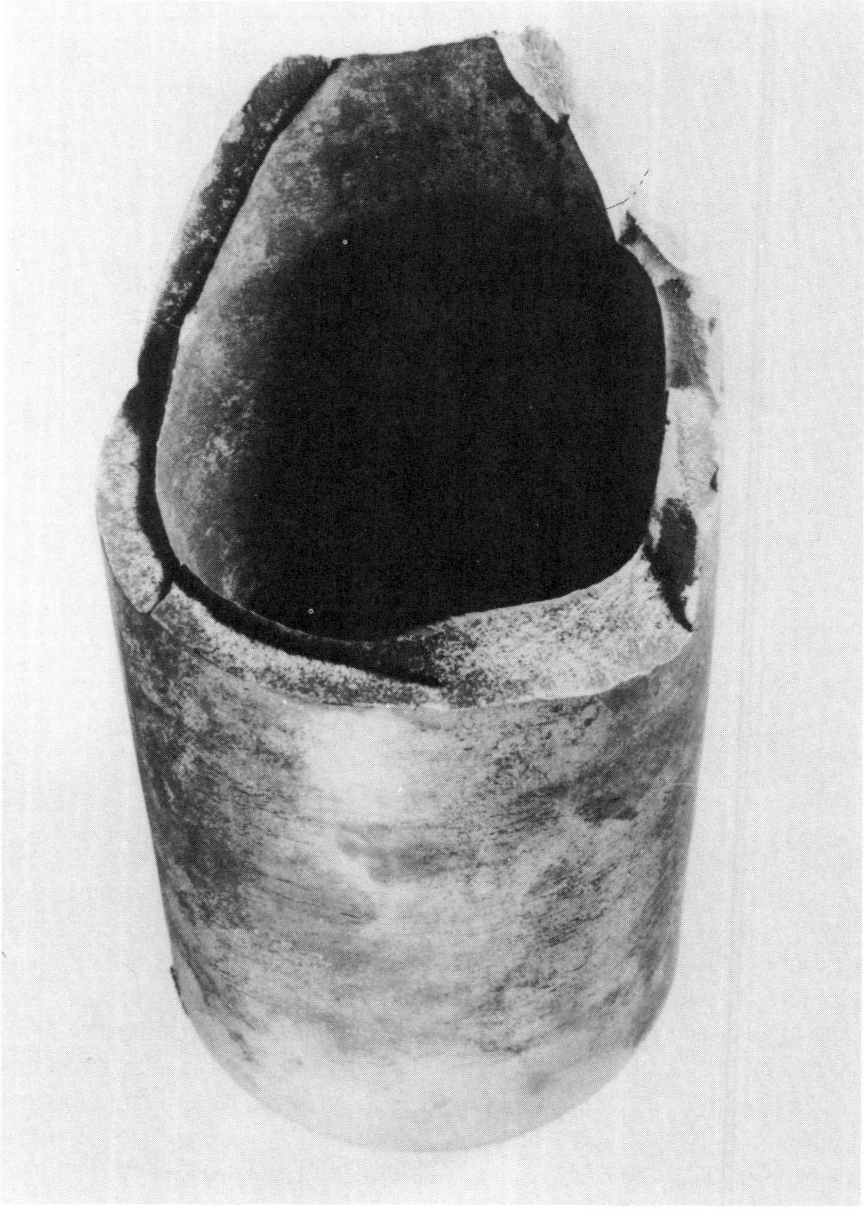

Figure 3–13 Sulfide-stress cracking failure of S-135 drillpipe (courtesy of G. Garwood).

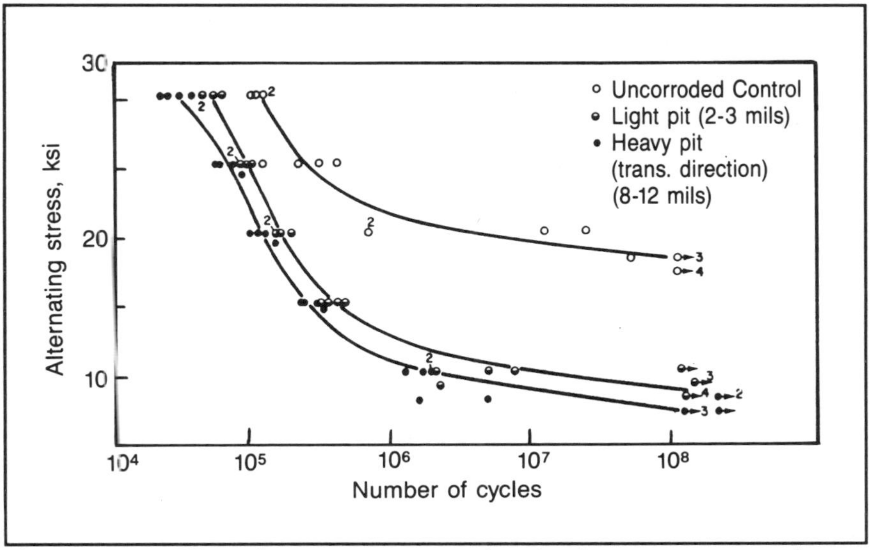

Figure 3–14 Effect of corrosion pitting on bending-fatigue properties of 2014-T6 sheet steel (after Person).

less frequently, they can be subjected to high hoop stresses; thus, cracking in the box typically occurs in a longitudinal direction. Drill collars fail in the same fashion as tool joints but less frequently since the applied stresses are lower from running on the bottom of the string.

Although hydrogen sulfide is most often derived from the reservoir, degradation of certain drilling muds can also produce hydrogen sulfide, thus contributing to SSC. Table 3–6 demonstrates the potential problems using lignosulfonate muds at high temperatures. Accompanying data to that in Table 3–6 by the same authors show that, even in high pH (>12.0) lignosulfonate mud, sulfide could be evolved because of decomposition of the mud (see Table 3–7). This decomposition has also been responsible for SSC failures of high strength casing when the mud was used as a packer fluid.

One metallurgical alternative to steel that has not gained wide use in the industry, despite certain advantages over carbon and low-alloy steel pipe, is aluminum drillpipe. One of its chief advantages is its lighter weight, which allows for deeper drilling with a mixed string. Although the yield strength (58,000 psi) and tensile strength (64,000 psi) of 2014-T6 aluminum are below that for Grade E drillpipe, these factors have been overcome by using thick wall sections. From a bending-fatigue standpoint, comparing steel and aluminum loaded to the same deflections, the aluminum pipe can have a fatigue life significantly greater than steel. Aluminum is also more resistant to H_2S and CO_2 corrosion than steel pipe.

Table 3–6 Thermal Degradation of Lignosulfonates at High Pressures.

Lignosulfonate Solution*	A	B	C
Temperature (°F)	450	405	425
Pressure (psi)	12,000	10,000	10,000
Initial pH	8.3	8.5	8.3
Final pH	6.1	5.9	5.9
H_2S (ppm)	15	2.5	20
Aging time (days)	16	16	16

*Solutions contained 10 parts per billion (ppb) lignosulfonate solubilized in a 0.43 ppb caustic soda solution. (After Bush et al.)

One of several disadvantages of aluminum drillpipe is its low resistance to muds with pH in excess of 10.5; increasing pH above this can cause serious corrosion. Chlorides can cause severe pitting, and as shown in Figure 3–14 can cause loss in fatigue resistance at levels greater than 180,000 ppm in salt muds. The endurance limit in this example has been reduced by more than 50%. Moreover, laboratory testing has confirmed that aluminum drillpipe is susceptible to stress-corrosion cracking in oxygenated brine muds even though in practice SCC has not been reported.

Another factor in the use of aluminum drillpipe is its behavior at elevated temperatures. Its fatigue strength is greatly reduced at temperatures of 300°F and greater (see Figure 3–15), so it is not recommended for long-term service above 250°F. However, this drawback is circumvented by using aluminum pipe in the top of the hole and steel pipe at the bottom of the hole, where the temperature is higher. Of course the static bottom-hole temperature is not the governing factor during drilling since mud flow will create a certain degree of cooling, so circulating temperature is an important consideration on temperature limitations. Figure 3–16 shows how the yield strength is reduced when tested at

Table 3–7 Thermal Degradation of Lignosulfonates in High-pH Solutions.

Lignosulfonate Solution*	A	B	C
Temperature (°F)	375	375	375
Pressure (psi)	300	300	300
Initial pH	12.6	12.8	12.6
Final pH	8.1	8.2	9.1
H_2S (ppm)	15	25	2.5
Aging time (days)	7	7	7

*Solutions contained 10 ppb lignosulfonate solubilized in 2 ppb caustic soda solution. (After Bush et al.)

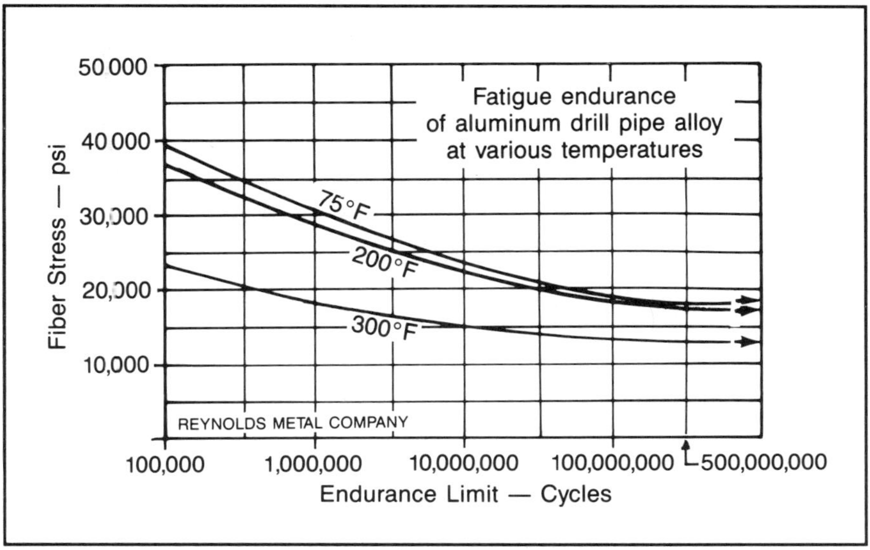

Figure 3–15 Fatigue strength of aluminum drillpipe at elevated temperatures (After Boice and Dalrymple, © 1963, SPE).

high temperatures and at room temperature after exposure to high temperature.

Another problem with aluminum pipe is its lower abrasion resistance compared to steel drillpipe. Because of its lower strength and therefore hardness, aluminum does not resist abrasion or erosion from drill cuttings in the mud as well as steel pipe. This problem can be alleviated somewhat by reducing the solids content of the mud; however, wear against casing can still be a problem.

Steel tool joints present another area of concern because of the galvanic corrosion problem they present and because their threaded joint has a tendency to unscrew during drilling. Both problems have been encountered in practice.

Titanium drillpipe represents another metallurgical alternative to steel and aluminum drillpipe. Titanium has been considered for very deep wells because of the prohibitively high hanging load for steel. Titanium offers a high-strength to low-weight ratio advantage over steel and it can be tool-jointed with titanium tool joints in contrast to aluminum drillpipe. Titanium is quite corrosion resistant to most drilling muds and the presence of H_2S and CO_2. Moreover, it has very good high-temperature strength for high bottomhole temperatures and excellent fatigue life.

One of the obvious disadvantages of titanium drillpipe compared to steel is the high cost. However, compared to the added cost of a large rig to handle steel pipe it may be competitive in some cases. Another disadvantage of titanium drillpipe is its higher wear rate against steel casing compared to steel drillpipe

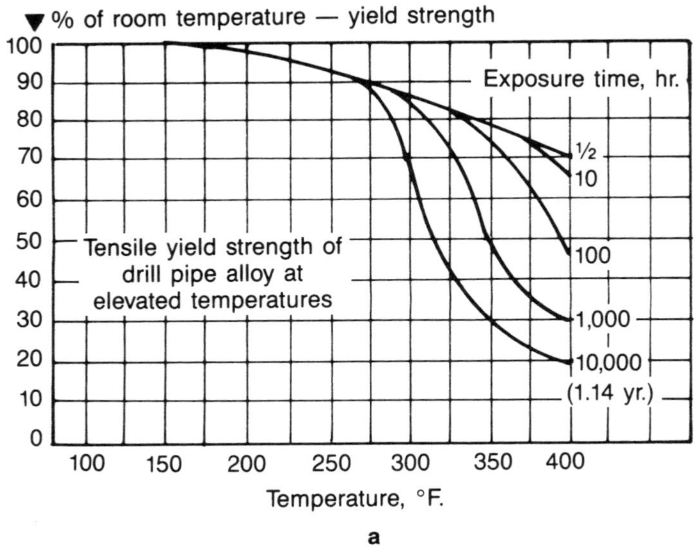

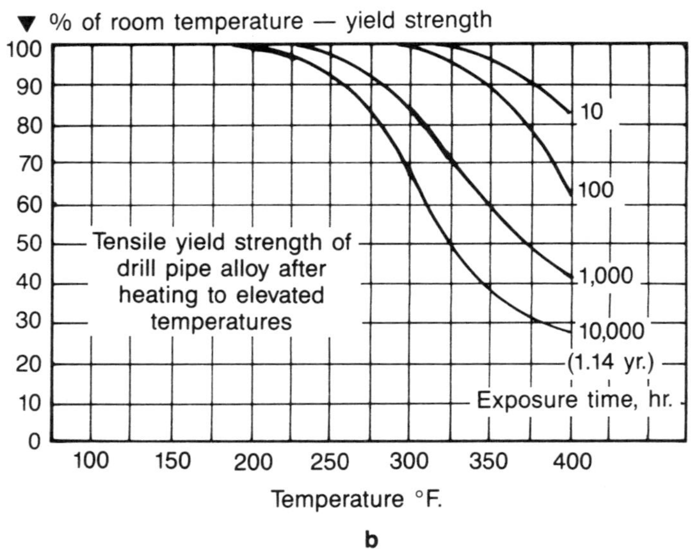

Figure 3–16 Yield strength of aluminum drillpipe (a) at elevated temperatures and (b) after exposure to elevated temperature (After McGhee).

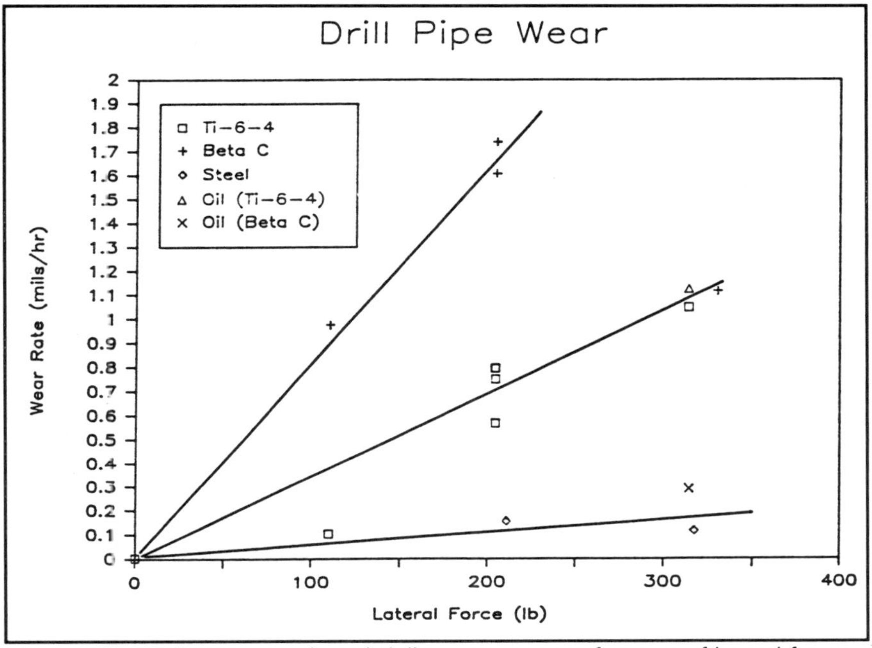

Figure 3–17 Titanium and steel drillpipe wear as a function of lateral force against P-110 casing in water-base and oil-base mud.

and tool joints. Figure 3–17 shows the wear rate of two types of titanium drillpipe (Ti-6-4 and Beta C) against P 110 steel casing compared to steel drillpipe in both water-base mud and oil-base mud. The mechanism is both corrosion and wear.

Tool joints most often suffer wear on the outside surface because of abrasion against the formation or casing during drilling in a deviated hole. Since the tool joint is larger in diameter than the pipe body, it has a greater tendency to rub the formation walls or casing than the body, although pipe-body failures have occurred when the pipe is crooked or when the deviation is so great that the pipe touches the walls. When tool joints are rotated under high lateral force against the hole wall, the temperature of the joint increases substantially because of friction. This temperature can be in excess of the lower critical temperature, eventually leading to heat checking of the tool joint as the joint is alternately quenched by the drilling mud and heated by friction. A network of longitudinal cracks form that can grow into longer cracks, offering sites for washout or further crack growth by fatigue or SSC.

The greatest reduction in torsional strength of a tool joint occurs by OD wear. This is caused both by lateral forces against the hole walls and by erosion from cuttings in the mud. Prevention of this type of problem is most often accom-

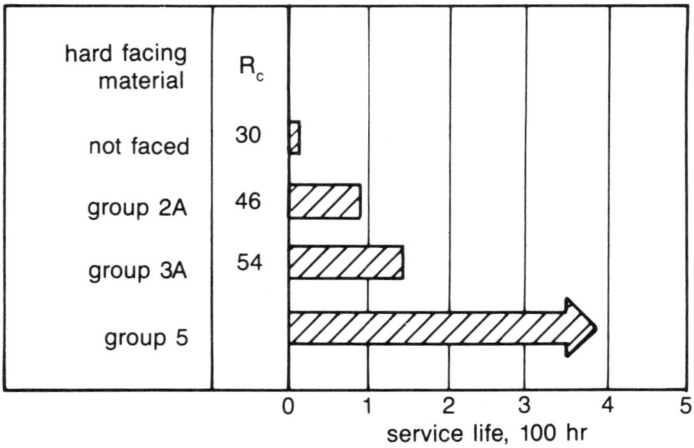

group 2A—12–15% chromium, molybdenum ferrous alloy
group 3A—25–50% chromium, nickel, molybdenum ferrous alloy
group 5—75–96% WC crystals in cobalt matrix

Figure 3–18 Wear resistance of various hard-facing alloys applied to tool joints of composition: 0.40% carbon, 1.5% manganese, and 0.2% molybdenum (After ASM, volume 1).

plished by hard facing (weld overlaying) the base metal with an alloy that has greater wear resistance. Some companies use gas metal arc welding, adding tungsten carbide to the weld puddle to enhance the wear resistance. The advantage of adding tungsten carbide for wear resistance is evident from the bar chart shown in Figure 3–18. Service life is significantly increased when tungsten carbide hard facing is used versus other hard-facing materials. Tungsten carbide is graded by particle size with fine mesh on the order of 325–400 mesh particles up to coarse mesh at about 8–10 mesh. While the coarse mesh has very good abrasion resistance, the individual particles tend to fracture more easily than the fine mesh due to the brittleness of tungsten carbide.

Figure 3–19 shows fracturing of large tungsten carbide particles in hard facing that resulted in failure of a drill collar. Similar failures may occur if the hard-faced piece is not properly preheated during welding and stress relieved after hard facing. Extremely high hardnesses are created in the heat-affected zone of the weld produced during the application of the hard face. If a post-weld heat treatment is not properly performed, premature failure of the drilling component will occur.

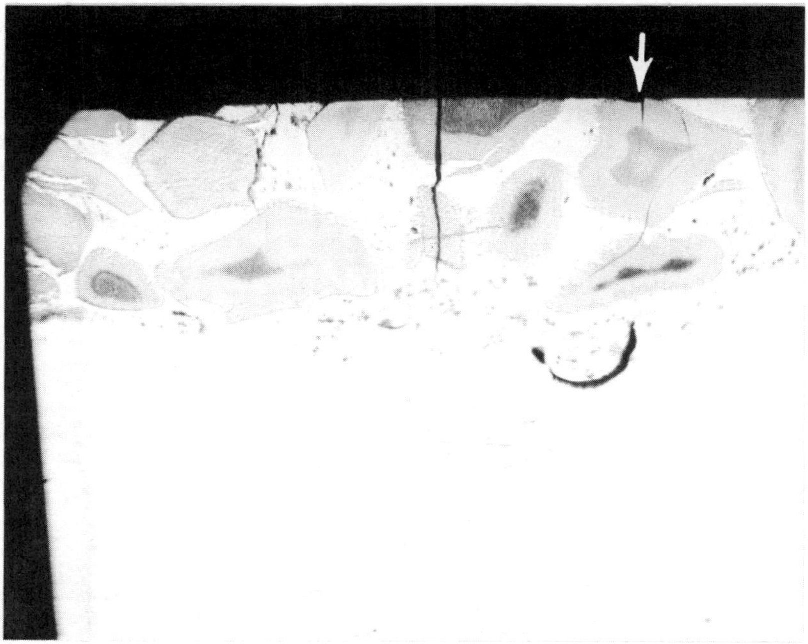

Figure 3–19 Cracks in tungsten carbide particles dispersed in hard facing on a drill collar.

Of course, one drawback of hardbanding is the increased wear on the casing. Casing wear must be a consideration when deciding on hard-faced or nonhard-faced tool joints. Figure 3–20 shows a comparison of wear rates on casing for coarse mesh, fine mesh, and nonhard-faced tool joints. Note that hard facing that has been ground smooth produces the least wear on casing.

Other failures of tool joints not associated with metallurgy are inadequate makeup torque leading to erosion of the pin or box threads by drilling mud, called wobble, and thread galling caused by forced makeup of improperly engaged threads.

Although casing will be discussed in the next chapter, one problem specific to drilling is fatigue of casing. As discussed, casing wear can be a considerable problem during drilling operations. However, another frequent problem encountered after long-term drilling through casing is fatigue of the casing connections. The threads acts as stress raisers and this coupled with the alternating load of the drill string rotating against the casing frequently produces a fatigue failure. Significant casing wear is not necessary for this event to occur. Figure 3–21 shows a fatigue crack in an eight-round casing thread connection in K-55 casing.

As should be evident, much effort still remains to improve the life of drilling

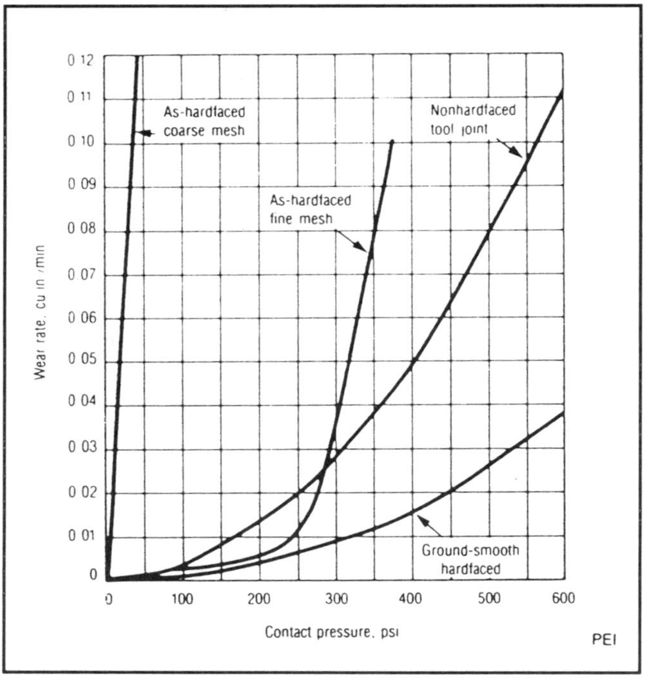

Figure 3–20 Casing wear rate as a function of contact pressure and type of hard facing (After Petroleum Eng. Int.).

components, especially the drillstem, in hostile drilling environments. Metallurgical considerations are of more importance in drilling today than ever before because of the deep drilling required to reach new reservoirs.

References

American Petroleum Institute. *Specifications 5D, 6A, 7, and 16A.*

American Petroleum Institute. *Recommended Practice 7G.*

American Society for Metals. *"Properties and Selection of Metals."* In *Metals Handbook.* Volume 1. 1961, p. 826.

Azar, J.J. "How O_2, CO_2, and Chlorides Affect Drill Pipe Fatigue." *Petroleum Engineer International* (March 1979), p. 72.

Boice, E.G. and R.S. Dalrymple. "The Design and Performance Characteristics of Aluminum Drill Pipe." *38th Annual Fall Meeting.* Paper 697. New Orleans, Louisiana: SPE of AIME, October, 1963, JPT.

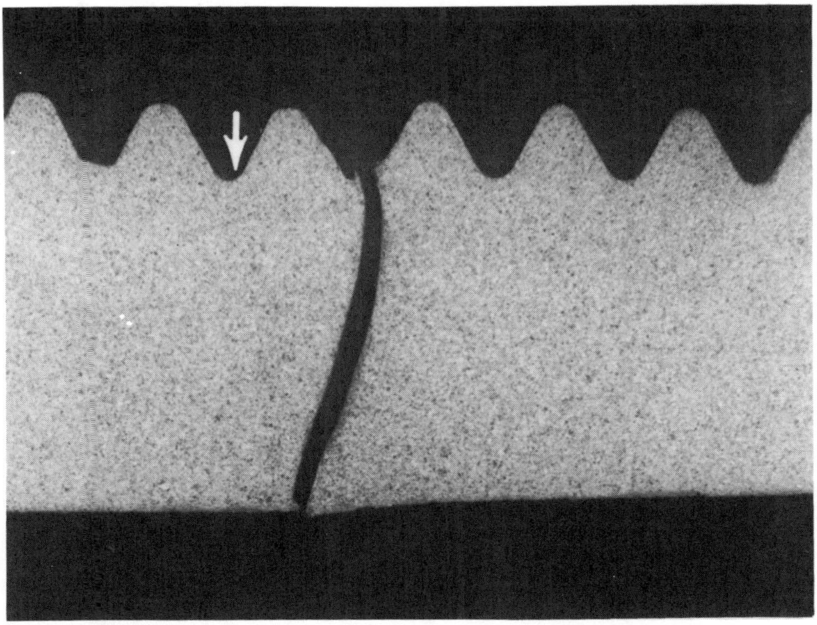

Figure 3–21 Cross section of 8-round casing threads showing progression of fatigue crack from thread root. Arrow indicates there are additional cracks in other thread roots.

Buscemi, C.D.; L. J. Klein; and G. B. Kohut. "Criterion Proposed to Reduce Drill Pipe Failures," *Oil and Gas Journal,* (Oct. 10, 1988), p. 64.

Bush, H.E.; R. Barbee; and J.P. Simpson. "Current Techniques for Combating Drill-Pipe Corrosion." *API Drilling and Production Practice.* 1966, p. 59.

Ferg, T.; B.D. Craig; and C. Aldrich, SPE Paper No. 23532 (1991).

Garwood, G.L. "Materials Selection for Downhole and Surface Equipment for Sour Gas Condensate Wells." *Corrosion/*73, Anaheim, California, 1973.

Grant, R.W. and H.T. Texter. "Causes and Prevention of Drill-Pipe and Tool-Joint Troubles." *API Drilling and Production Practice.* 1941, p. 9.

McGhee, E. "What You Should Know About Aluminum Drill Pipe." *Oil and Gas Journal* (March 18, 1963), p. 168.

McGlasson, R.L. "Special Metallurgical Problems." *Proceedings of the University of Oklahoma Corrosion Control Course,* Norman, Oklahoma, September 1971.

Mehdizadeh, P. "Drill Pipe Failures: Where Do We Go From Here." *Petroleum Engineer* (September 1973), p. 52.

Person, N.L. "Fatigue Properties of Prior-Corroded Aluminum Sheet Alloys." *Material Performance,* 14 (1975), p. 22.

Petroleum Engineer International, p. 18 (March 15, 1984).

Rollins, H.M. "What Do We Know About Drill Pipe Fatigue Failure." *Oil and Gas Journal* (April 18, 1966), p. 98.

Rollins, H.M. "Drill-Stem Failures Due to H_2S." *Oil and Gas Journal* (January 24, 1966), p. 82.

"Steels for Conventional Energy Production and Transmission." *Metal Progress,* 108 (1975), p. 59.

Tsukano, Y.; H. Miyoshi; Y. Sogo; S. Nishi; and E. Takeuchi. "Optimum Design of Internal Upset Geometry of Drill Pipe for Longer Fatigue Life." Nippon Steel Technical Report No. 44, p. 37 (Jan. 1990).

Tuttle, R.N. "Deep Drilling—A Materials Engineering Challenge." *Materials Performance,* 13 (1974), p. 42.

Williamson, J.S. and Bolton, J.B. *Petroleum Engineer International,* (September 1983), p. 54 and *Petroleum Engineer International,* (March 15, 1984), p. 18.

4

METALLURGY FOR OIL AND GAS PRODUCTION

The components discussed in this chapter are casing, tubing, sucker rods and pumps, Christmas tree components, and wire lines. As in the case of drilling, there are a multitude of special tools and equipment for oil and gas production, many of which have proprietary metallurgy. Although these specialty tools are important to the industry, they do not represent the same capital expenditure as tubing, casing, valves, etc., or the same potential for causing a catastrophic failure.

Production equipment receives more attention from a research and development standpoint than other types such as drilling equipment simply because the life of a well may be projected to 20 years or greater in some instances. For such exceptionally long service under severe conditions, extraordinary measures to combat corrosion are often called for, including the application of corrosion resistant alloys. It is the recent progression and advancement of the industry from primarily carbon-steel components to high-alloy steels, stainless steels, nickel-base alloys, and titanium alloys that will be discussed in this chapter as well as the reasons for the necessity of such a transition.

Casing and Tubing

Casing, tubing, and drillpipe or, as they are often referred to in the industry, *oil-country tubular goods* (OCTG) are governed by the API under *Specifications*

Table 4–1 Chemical Requirements (By Weight Cent)

0 Group	1 Grade	2 Type	3 Carbon min.	3 Carbon max.	4 Manganese min.	4 Manganese max.	5 Molybedenum min.	5 Molybedenum max.	6 Chromium min.	6 Chromium max.	7 Nickel max.	8 Copper max.	9 Phos-phorous max.	10 Sulfur max.	11 Silicon max.
1	H-40	...	...	...	...	...	...	...	...	...	...	...	0.030	0.030	...
1	J-55	...	...	...	...	...	...	...	...	...	...	...	0.030	0.030	...
1	K-55	...	...	...	...	...	...	...	...	...	...	...	0.030	0.030	...
1	N-80	...	...	...	...	...	...	...	...	...	...	...	0.030	0.030	...
2	L-80	1	...	0.43[1]	...	1.90	...	...	...	...	0.25	0.35	0.030	0.030	0.45
2	L-80	9Cr	...	0.15	0.30	0.60	0.90	1.10	8.00	10.00	0.50	0.25	0.020	0.010	1.00
2	L-80	13Cr	0.15	0.22	0.25	1.00	...	...	12.0	14.0	0.50	0.25	0.020	0.010	1.00
2	C-90	1	...	0.35	...	1.00	0.25[2]	0.75	...	1.20	0.99	...	0.020	0.010	...
2	C-90	2	...	0.50	...	1.90	...	NL	...	NL	0.99	...	0.030	0.010	...
2	C-95	...	...	0.45[3]	...	1.90	...	...	...	...	...	...	0.030	0.030	0.45
2	T-95	1	...	0.35	...	1.20	0.25[4]	0.85	0.40	1.50	0.99	...	0.020	0.010	...
2	T-95	2	...	0.50	...	1.90	...	...	...	...	0.99	...	0.030	0.010	...
3	P-110	...	...	...	...	...	...	...	...	...	...	...	0.030	0.030	...
4	Q-125	1	...	0.35	...	1.00	...	0.75	...	1.20	0.99	...	0.020	0.010	...
4	Q-125	2	...	0.35	...	1.00	...	N.L.	...	N.L.	0.99	...	0.020	0.020	...
4	Q-125	3	...	0.50	...	1.90	...	N.L.	...	N.L.	0.99	...	0.030	0.010	...
4	Q-125	4	...	0.35	...	1.90	...	N.L.	...	N.L.	0.99	...	0.030	0.020	...

[1]The carbon content for L-80 may be increased to 0.50% max. if the product is oil quenched.
[2]The molybdenum content for Grade C-90. Type 1 has no minimun tolerance if the wall thickness is less than 0.700 inch.
[3]The carbon content for C-95 may be increased to 0.55% max. if the product is oil quenched.
[4] The molybdenum content for Grade T-95. Type 1 may be decreased to 0.15% minimum if the wall thickness is less than 0.700 inch
N.L. = No Limit. Elements shown must be reproted in product analysis.
(courtesy API)

5CT and *5D*. The various API grades with their corresponding chemistries and mechanical properties are shown in Tables 4–1 and 4–2, respectively. The Group One grades represent the lowest strength, and the Group Four, the highest-strength grades. The specifications require manufacture of the steel by electric furnace, open hearth, or the basic oxygen process. The chemistry requirements are minimal for Groups One and Three, with only the phosphorus and sulfur restricted to a maximum of 0.030 wt% each. More control on chemistry is provided in Groups Two and Four.

Pup joints, casing, and tubing made to this specification can be either seamless or electric-resistance welded (ERW) depending on the particular grade. The ERW casing and tubing has to be either heat-treated after welding to a minimum temperature of 1,000°F (538°C) or processed so there is no untempered martensite. This latter requirement by API 5CT has created problems with some ERW that is not heat- treated after welding since it is quite difficult to determine that no untempered martensite is present.

API 5CT only requires a weld-seam heat treatment. The seam is usually induction heated to about 1,650°–1,850°F for less than 1 min, thus producing a normalized structure similar to the pipe body. The time at temperature (1,650°–1850°F) is short enough that grain growth is not a problem. In fact, the induction heat treatment is so fast, the lower critical temperature may not be surpassed and the weld only receives a temper.

Grade N-80 tubing and work tubing are required to be heat-treated full length and, if upset, re-heat-treated full length after upsetting. Grades J, K, and N casing and Grade J tubing are only required to be normalized if specified by the purchaser; otherwise they are supplied in the as-rolled condition. Grade J-55 tubing is typically not normalized full length after upsetting, which has been the source of numerous failures in CO_2 service over the years. This type of corrosion is referred to as ringworm corrosion and will be discussed later in this chapter. For the high-strength Groups Three and Four, the pipe must be either quenched and tempered or normalized and tempered and, if upset, full-length heat-treated after upsetting.

Couplings are required to be seamless and of the same grade as the pipe, meeting the mechanical requirement set forth in Table 4–2. However, Grades H-40 and J-55 pipe can be furnished with either J-55 or K-55 couplings.

Grades covered by Group Two have controlled chemistry and mechanical properties to provide greater resistance to sulfide-stress cracking. The chemistry is considerably more restricted than for the other groups and also includes 9 Cr and 13 Cr compositions. The range of yield strength is much tighter for these grades such as L-80 versus N-80, the first of which has an allowable yield strength spread of 15,000 psi and the latter of 30,000 psi. Furthermore, the Group Two grades have hardness maxima and ranges of allowable hardness variations.

Table 4–2 Tensile and Hardness Requirements

1	2	3	4		5		6		7		8	9
			Yield Strength				Tensile Strength Min.		Hardness Max.*			Allowance
Group	Grade	Type	Min.		Max.						Specified Wall Thickness Inches	Hardness Variation. HRC
			psi	MPa	psi	MPa	psi	MPa	HRC	BHN		
	H-40		40,000	276	80,000	552	60,000	414	...	...		
	J-55		55,000	379	80,000	552	75,000	517	...	...		
1	K-55		55,000	379	80,000	552	95,000	655	...	...		
	N-80		80,000	552	110,000	758	100,000	689	...	...		
	L-80	1	80,000	552	95,000	655	95,000	655	23	241		
	L-80	9Cr	80,000	552	95,000	655	95,000	655	23	241		
	L-80	13Cr	80,000	552	95,000	655	95,000	655	23	241		
	C-90	1.2	90,000	620	105,000	724	100,000	690	25.4	255	0.500 or less	3.0
	C-90	1.2	90,000	620	105,000	724	100,000	690	25.4	255	0.501 to 0.749	4.0
2	C-90	1.2	90,000	620	105,000	724	100,000	690	25.4	255	0.750 to 0.999	5.0
	C-90	1.2	90,000	620	105,000	758	100,000	690	25.4	255	1.000 and above	6.0
	C-95		95,000	655	110,000	758	105,000	724	...	...		
	T-95	1.2	95,000	655	110,000	758	105,000	724	25.4	255	0.500 or less	3.0
	T-95	1.2	95,000	655	110,000	758	105,000	724	25.4	255	0.501 to 0.749	4.0
	T-95	1.2	95,000	655	110,000	758	105,000	724	25.4	255	0.750 to 0.999	5.0
3	P-110		110,000	758	140,000	965	125,000	862	...	...		
	Q-125		125,000	860	150,000	1035	135,000	930	...	...	0.500 or less	3.0
4	Q-125		125,000	860	150,000	1035	135,000	930	...	...	0.501 to 0.749	4.0
	Q-125		125,000	860	150,000	1035	135,000	930	...	...	0.750 and above	5.0

*In case of dispute, laboratory Rockwell C hardness tests shall be used as the referee method.
(courtesy API)

Heat treatment is also more detailed for Group Two. Heat treatment after upsetting and full-length heat treatment as in the other groups are required, along with minimum tempering temperatures. Cold working after the final tempering treatment is forbidden except for C-95 pipe, which is allowed up to a maximum of 3% compressive cold working after tempering.

Couplings are required to be seamless and of the same grade as the pipe, meeting all of the same chemical composition and mechanical properties as the pipe.

J-55 and K-55 are the most popular grades for tubing and casing, especially for shallow wells. Both usually are relatively high-carbon (0.45%) steels with a chemistry similar to AISI 1045 steel. They are provided in the normalized or as-rolled condition but, because of the high carbon content, can reach strengths on the high end of the API requirement of 80,000 psi (552 MPa) and thus be equivalent to N-80 grade. H-40 grade is less frequently used; and with deeper wells requiring higher strength, burst, and collapse, its application has diminished in recent years.

Grades C-75, N-80, L-80, P-105, and P-110 are typically produced from C-Mn or C–Mn–Mo steels. The C-75 and P-105 grades were earlier API grades that are no longer part of the specification. C-90, T-95 and C-95 are made from steels based on 4130 steel and commonly modified with niobium, molybdenum, titanium, and/or boron.

Coupling stock often has the same general chemistry as the pipe but may contain slightly greater concentrations of molybdenum, titanium, boron, or manganese to increase hardenability since couplings have thicker walls than the pipe and must be able to achieve the same strength levels as the body.

Many non-API grades of tubulars are available for special applications. These range from high collapse strength casing, to low-alloy steels with restricted properties for hydrogen sulfide service, to nickel-base and cobalt-base alloys for deep, hot wells containing chlorides, hydrogen sulfide, and carbon dioxide. These latter alloys are referred to as corrosion-resistant alloys (CRA). Some of the more common of these alloys will be discussed later.

At this point it is appropriate to briefly discuss seamless and ERW pipe. There are advantages and disadvantages in using either type, depending on the application. Any time a weld is introduced into a structure, it becomes a better candidate for failure than the base material, which has a more homogeneous structure. The weld area by virtue of the welding process has a greater propensity for defects and, thus, a higher probability of failure than wrought base metal. However, properly manufactured and well-inspected ERW pipe has provided excellent service to the industry over many years and can be comparable to seamless. The key is careful control of the manufacturing of ERW and 100% inspection of the weldline.

Figure 4–1a Split along ERW seam in tubing caused by internal pressure and poor weld. Note stitching on the left side of the split.

Weld defects in ERW pipe arise from several sources. Foreign matter on the welding surfaces, such as oxides, short-circuiting the welding current, arcing electrical contacts, and excessive or insufficient upsetting during welding will produce regions of no weld. In the past many of these problems were the result of using low frequency welding. Now with the widespread use of high frequency welding (HFW) ERW quality has been greatly improved. Figure 4–1a is an example of ERW tubing that was intermittently welded (stitching). It shows the stitching appearance in the enlargement (Figure 4–1b). Another common weld defect is hook cracks. These cracks form along the flow lines of the material that is upset during welding and are associated with nonmetallic inclusions that are deformed during upsetting. Figure 4–2 shows the cross-section of a weld with a hook crack originating at an inclusion.

Other problems with the ERW seam may arise during heat treatment after welding. Although the API requires a minimum heat treatment of 1,000°F (538°C) or processing in a manner such that no untempered martensite remains, hard spots can still be present. This, of course, reduces fracture toughness as well as SSC resistance. Also, during heat treatment of the weld, sufficient heat may not fully penetrate the wall thickness or misalignment of the induction heat zone with respect to the weld may occur. Unheat-treated welds can be quite hard and brittle providing an easy fracture path for failure that is not detectable until the pipe is run in the field. Figure 4–2 also shows the hourglass pattern of the heat-affected zone which is characteristic of an ERW that had not been heat-treated. Figure 4–3 shows a weld that was properly heat-treated, leaving only the tell-tale white ferrite line.

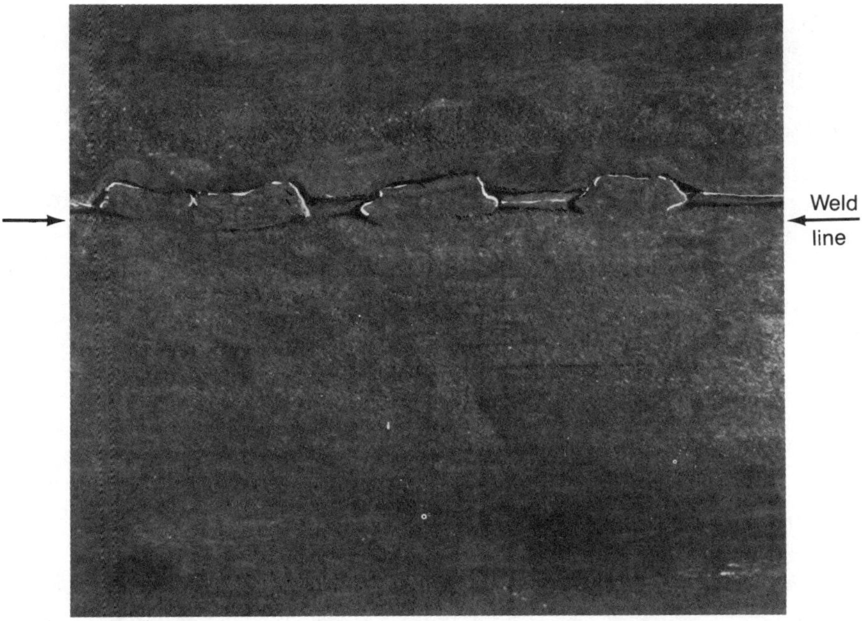

Figure 4–1b Enlargement of left-hand portion of weld in Figure 4–1a.

Improved manufacturing and better inspection techniques in the past few years have significantly reduced problems encountered with ERW pipe, making it competitive with seamless. Furthermore, ERW pipe has certain advantages over seamless. The surface finish is typically much better for ERW than seamless; thus coating application is less troublesome and less expensive on ERW pipe. The wall thickness and concentricity of ERW pipe is much more uniform than seamless, giving it better collapse resistance than seamless. Finally, it is less expensive than seamless pipe. However, the variability in quality from manufacturer to manufacturer is great and care should be taken in selecting the specific ERW to be used. For a majority of casing and tubing applications in the oilfield, ERW pipe is more than satisfactory if it has been properly manufactured, inspected, and tested.

The comments on ERW pipe should not be taken to imply there are no potential problems with seamless pipe. In fact, seamless pipe also suffers from manufacturing problems such as seams, scabs, cracks, etc. Figure 4–4 is an example of Q-125 seamless casing that fractured as a result of a seam that formed in the casing during hot working after piercing. Moreover, the surface condition of seamless can be quite poor as a result of hot rolling during manufacturing, thereby adversely impacting the application of coatings.

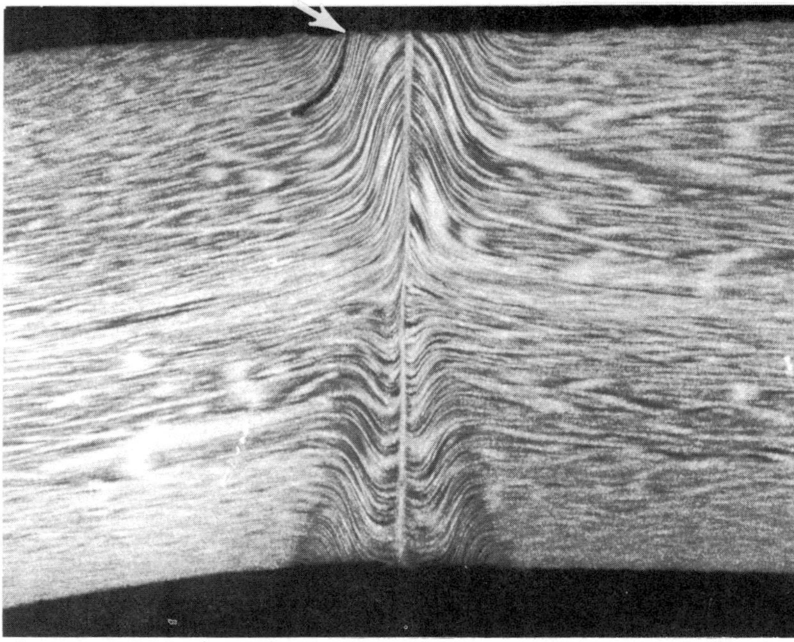

Figure 4–2 Hook crack (arrow) in ERW Casing. The vertical white line at the center is the weld. Mag. 12X.

Regardless of whether the pipe is seamless or ERW, it can fail by modes of loading that are encountered in well operations. It is the requirement of the casing or tubing string designer to take into account tensile loading, burst, collapse, and connection jumpout. All of these modes are addressed by API bulletins and specifications. However, improper design or changes in the well environment can cause failures from these factors. Generally, failure from any one of these loading conditions are easy to identify.

Figure 4–5 shows the characteristic features of a tensile overload. Prior to complete fracturing at the center, the pipe OD and ID substantially neck down, displaying a high degree of ductility. This type of failure occurs simply from pulling the pipe apart.

A collapse failure also has very obvious features, Figure 4–6. In collapse, the pressure on the outside of the pipe is higher than inside so it collapses inward. Not all collapses are as complete as this and frequently just show a partial collapse of the pipe.

Burst is just the reverse of collapse with pressure on the inside of the pipe causing it to split open. The split is typically in the longitudinal direction.

Connection jumpout is one of the best understood and most common but least

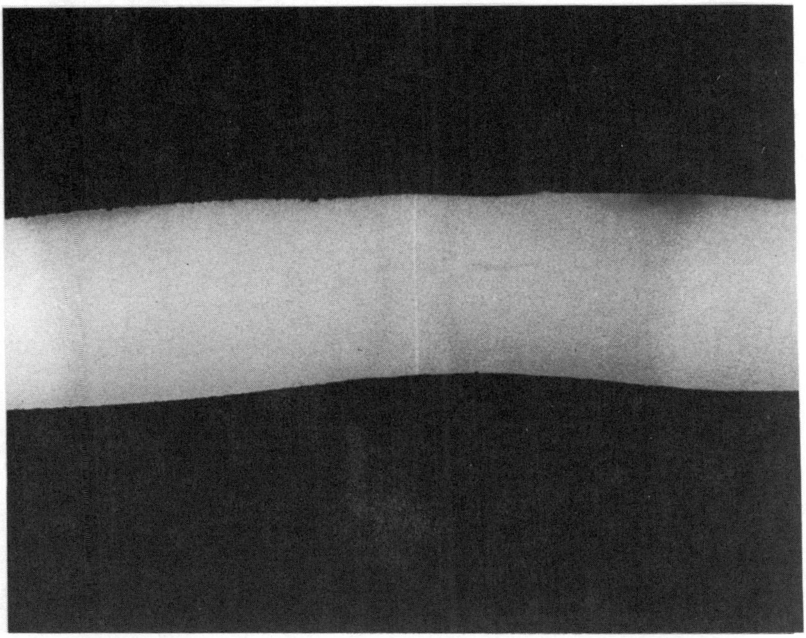

Figure 4–3 Cross section of properly heat-treated ERW displaying only the white ferrite line.

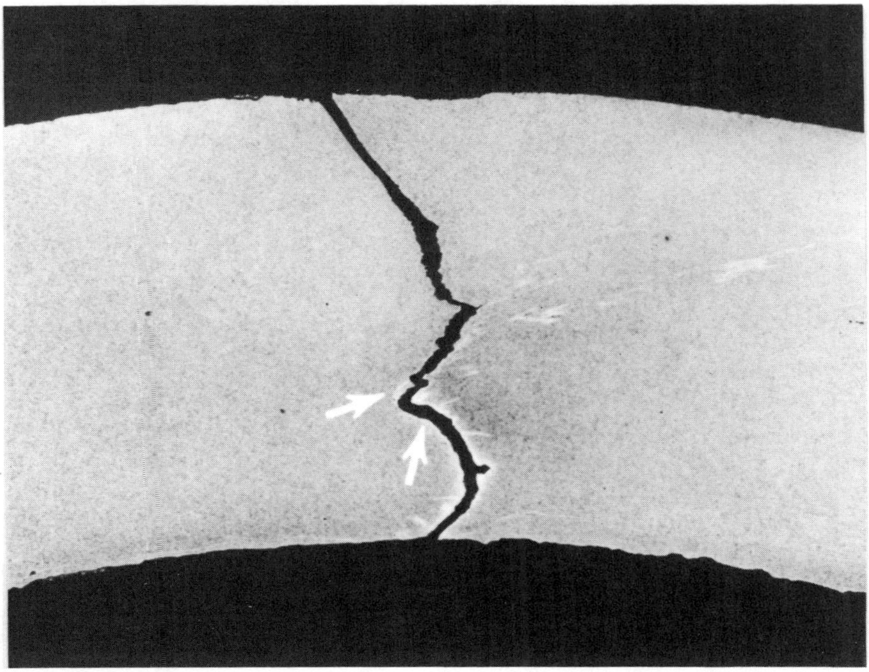

Figure 4–4 Macrograph of a section through the fracture origin of Q–125 seamless casing. Arrows indicate original seam.

Figure 4–5 Tensile overload of P–105 tubing showing the necking-down of the pipe in the vicinity of the fracture. Mag. 3/4X.

recognized problems in the industry. Figure 4–7 shows the undamaged pin threads next to the coupling of a 5-1/2-in., 15.5 pound per foot, K-55, 8-round connection. On closer examination, the thread dimensions, tensile properties, and overall dimension of the pin and coupling were in accordance with the API

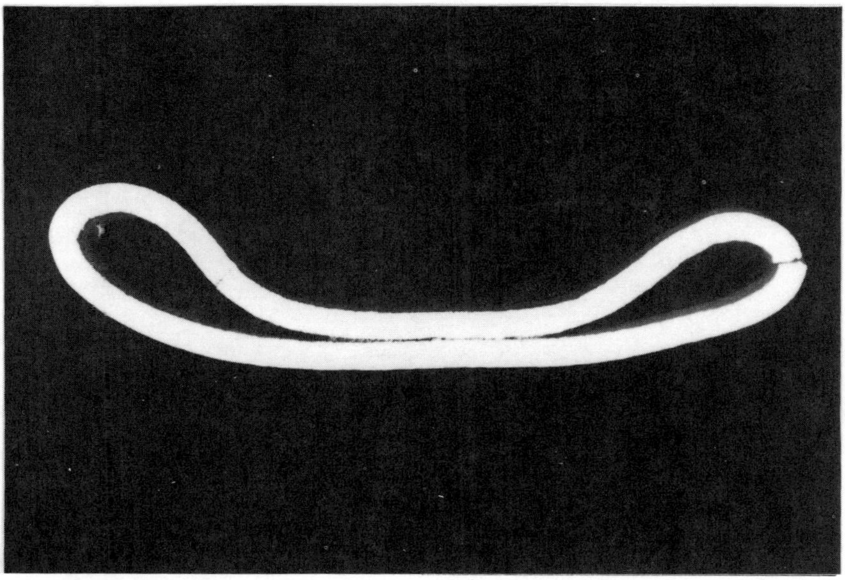

Figure 4–6 End view of a collapsed tubing section. Mag. 1X.

requirements. Excessive tensile loading is the most frequent cause for thread jumpout although the other factors such as a soft pin, and incorrect thread dimensions can also induce jumpout.

During heat treatment of casing and tubing, specifically during the quenching of steels, if the quench is too rapid or the steel too high in carbon or alloying elements, the pipe can crack. Quench cracking results when outer regions of the steel transform to martensite before inner areas in the workpiece do, such that stresses from the transformation of austenite to martensite cannot be accommodated and cracking results. Figures 4–8a and 4–8b show quench cracks in a C-140 grade casing material. Quench cracking has been observed in almost all API grades of casing and tubing that is quenched and tempered including L-80 13 Cr. Quench cracking usually originates at the OD since it is first to transform and the cracking is always intergranular.

Another consideration in casing and tubing is fracture toughness. In the past this was not an important consideration because low strength pipe (e.g., J-55, K-55, etc.) was sufficiently tough at wellbore temperatures that fracture was not a problem. However, with the increasing use of these grades at arctic temperatures and the need to use casing with high collapse resistance, which translates to high strength, fracture toughness has become an important factor. Moreover, if the tubing or casing is not sufficiently tough then it may crack during normal well operations, for example, from the introduction of cold frac fluids, acid, etc.

Figure 4–7 Pin and coupling after jumpout of the pin. There was no damage to either the pin or coupling threads. Mag. 1/3X.

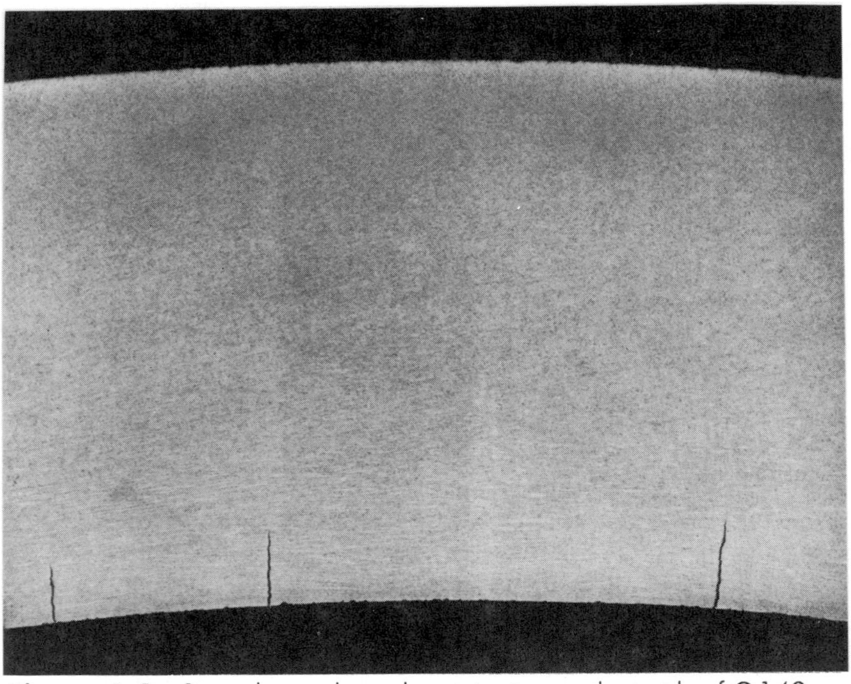

Figure 4–8a Several quench cracks originating on the inside of C-140 casing.

API 5CT recognized this need by requiring Charpy impact testing of Q-125 grade casing and specific minimum impact energies depending on wall thickness. At a very minimum, higher strength proprietary casing grades should also be impact tested to ensure adequate toughness for well control.

Figure 4–9 presents Charpy impact data for a typical K-55 casing as hot-rolled. The fracture toughness is quite poor, displaying a ductile-to-brittle transition temperature well in excess of ambient. Normalizing has been shown to greatly improve toughness over the as-rolled condition. However, J and K are most often supplied in the latter condition.

While purely mechanical failures of tubulars such as collapse, burst, tensile failure, etc., occur, they are less common than corrosion. Until recently, corrosion in petroleum production was always handled by using plastic coatings and/

Figure 4–8b One crack at higher magnification showing the intergranular nature, 400X.

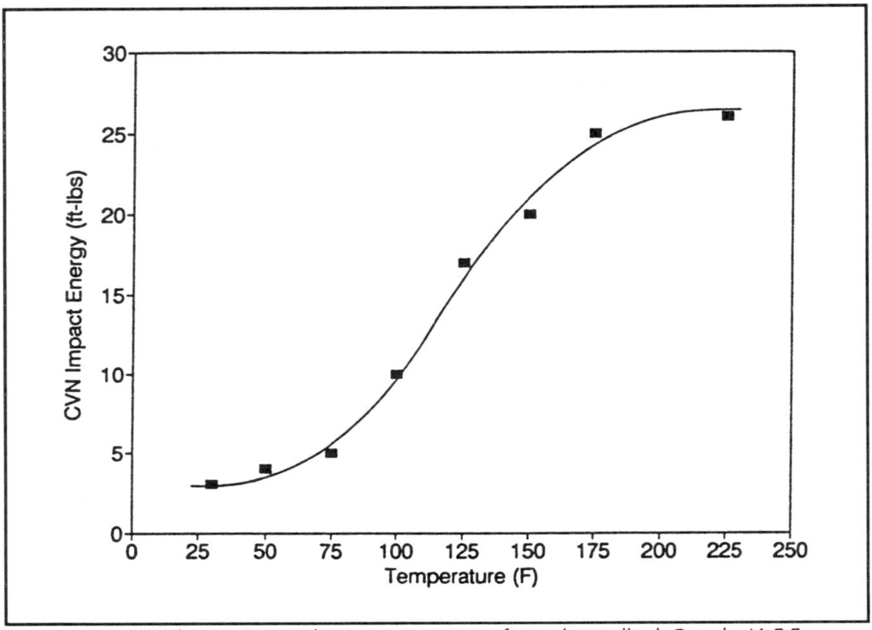

Figure 4–9 Charpy V-notch transition curve for a hot-rolled Grade K-55 casing.

or chemical inhibitors. However, with increasingly hostile well environments, the ability of coatings and inhibitors to withstand temperatures exceeding 375°F and pressures of 10,000 psi has been severely reduced. The only effective alternative for corrosion mitigation that remains is metallurgy. Although 9 Ni and 9 Cr steels had been used for tubing in the early 1950s, no further development occurred until the deep, high-hydrogen sulfide wells in Mississippi required alternative solutions to the standard coating/inhibitor program.

Produced oil and gas often includes corrodents such as hydrogen sulfide, carbon dioxide, and chlorides. It is quite rare in the production of these fluids that water is not present in at least sufficient amounts to cause corrosion. Pitting by hydrogen sulfide, carbon dioxide, and chlorides can become severe, and, when combined with high producing temperatures, corrosion rates on the order of several hundred mils per year (mpy) to an inch per year (ipy) have been observed. In addition to pitting attack, more catastrophic forms of failure such as sulfide-stress cracking (SSC) and in the case of stainless-steel and nickel alloy tubulars, stress-corrosion cracking (SCC) can occur.

Only those forms of corrosion attack that are metallurgically significant will

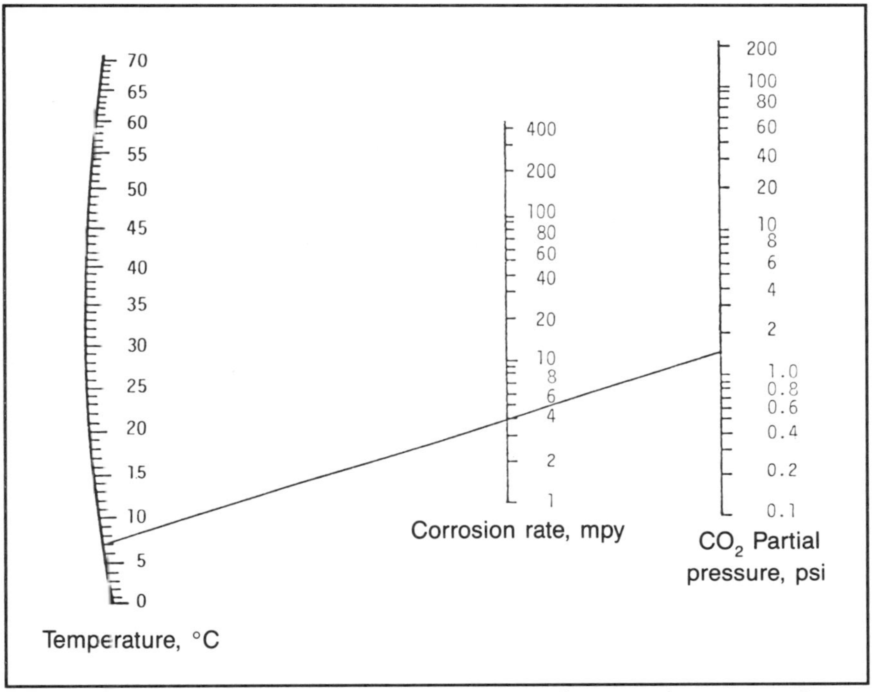

Figure 4–10 Calculation of corrosion rates as a function of CO_2 partial pressure and temperature. For example, at 7°C and 1.5 psi CO_2 partial pressure, the corrosion rate is 4 mpy (After DeWaard and Milliams, © BHRA).

be discussed. These are forms that can be affected by metallurgical solutions such as pitting, SSC, stress-corrosion cracking and general weight-loss corrosion.

Inasmuch as possible, inhibitors are still the primary method for corrosion control in producing wells. When this approach is unsatisfactory or ineffective, the next step is an increase in alloy content of the tubulars. Although there has been some exposure of 9 Cr-1 Mo tubulars to aggressive environments, the resistance of this alloy is marginal. It is more common to use a stainless-steel grade such as AISI 420, which contains nominally 13% Cr. Carbon dioxide in the produced gas or oil can lead to very rapid deterioration of alloys containing less than 12% Cr. DeWaard and Milliams described the severe weight-loss corrosion caused by carbon dioxide and presented their findings as well as those of others in the nomograph shown in Figure 4–10. Corrosion caused by carbon dioxide is primarily a function of CO_2 partial pressure and temperature. This is more apparent from the equation developed by DeWaard and Milliams:

$$\text{Log } r = 8.78 - \frac{2{,}320}{(T + 273)} - 5.55 \times 10^{-3}T + 0.67 \text{ Log } Pco_2$$

where:

r	=	corrosion rate, mpy
T	=	temperature, °C
Pco_2	=	partial pressure of CO_2, psia

For example, assuming a partial pressure for carbon dioxide of 1.5 psi at a temperature of 7°C, the corrosion rate would be 4 mpy. The equation is useful for those cases that extend beyond the nomograph although it is not strictly valid for conditions beyond 15 psia CO_2 partial pressure. However, it has been found to apply under some conditions and is useful for illustrating the potential severity of CO_2 corrosion. Thus for a deep well with a 20,000-psi bottom-hole pressure and 10% CO_2, the partial pressure would be about 2,000 psi. If the bottom-hole temperature were 176°C (350°F), the calculated corrosion rate would be 70.5 ipy (70,500 mpy). This value is unreasonably high and shows one of the limitations of the DeWaard and Milliams approach. The equation does not consider the very important corrosion rate limiting effect of the formation of a protective or semi-protective iron carbonate ($FeCO_3$) layer on the steel. Nor does it take into account the produced water composition (e.g., calcium and bicarbonate content) that can have a profound affect on reducing the CO_2 corrosion rate. Nevertheless, extremely high corrosion rates have been found in fields operating with high carbon dioxide partial pressures and high temperatures. The nomograph and equation still provide a means to approximate the severity of carbon dioxide corrosion and allow a quick, albeit conservative (worst case), analysis of the potential corrosive behavior of a carbon dioxide-containing system. Also, this method gives at least some semiquantitative information at low partial pressures of CO_2 versus the old API rule of thumb that, for carbon dioxide partial pressures below 7, no corrosion occurs; between 7–30, moderate corrosion can be expected; and above 30, corrosion will be severe.

The work of DeWaard and Milliams covered carbon dioxide partial pressure only to 1.0 bar. Work by others has extended the study of CO_2 corrosion above 1.0 bar, and their results are shown in Figures 4–11a and 4–11b. Since temperature and carbon dioxide partial pressure change with location in the tubing string, it is obvious the corrosion behavior will vary along the length of the string.

Figure 4–12 shows an example of CO_2 corrosion of an oil well tubing string. Corrosion by carbon dioxide, which forms carbonic acid when combined with water, is particularly severe in threads, connections, changes in metallurgical

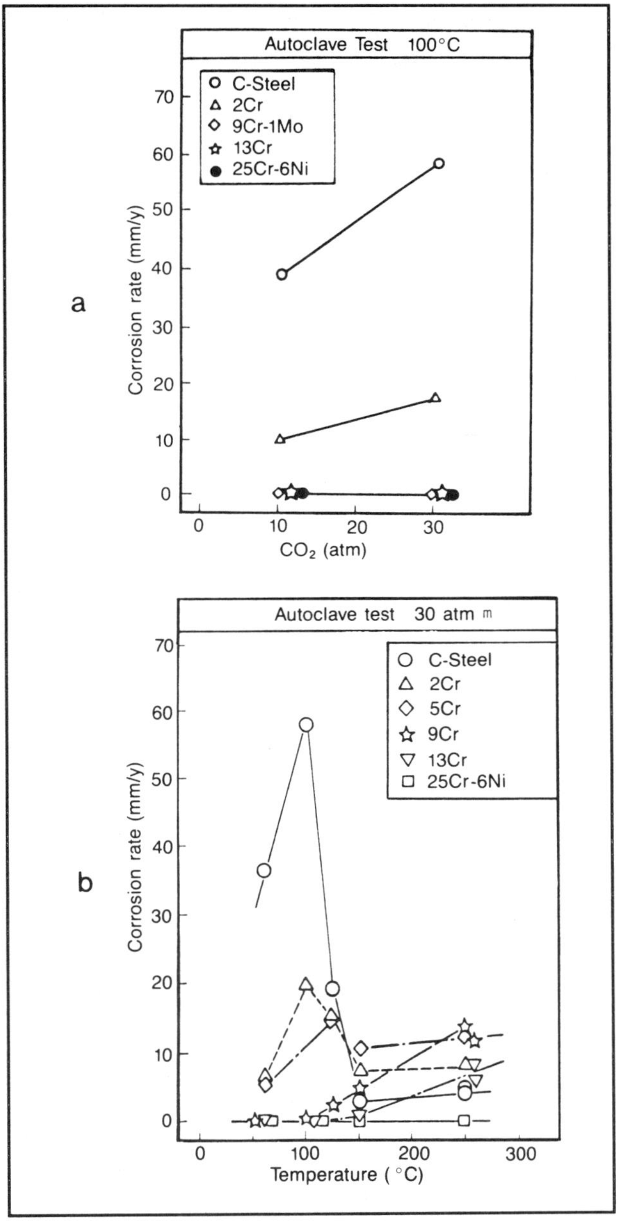

Figure 4-11 Effect of CO_2 Gas partial pressure on corrosion of chromium steels (a) and effect of temperature on CO_2 corrosion rate of various chromium steels (b) (After Sumitomo).

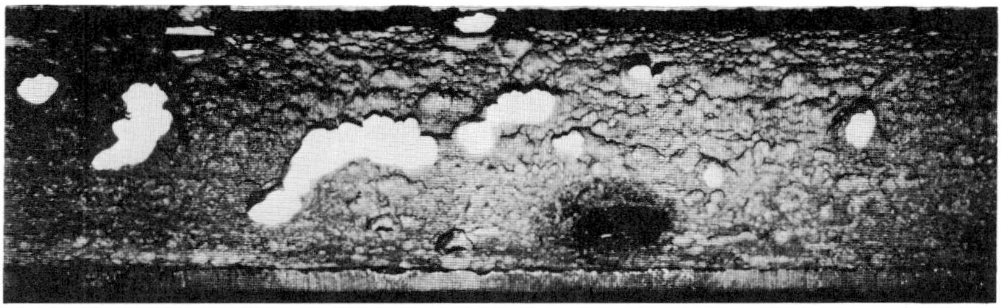

Figure 4–12 CO_2 corrosion of oil well tubing.

structure (e.g., welds and upsets), and changes in flow such as through connections and valves. Severe problems have been observed in gas production containing carbon dioxide when tubing and couplings of different grades are mixed. N-80 couplings connected to J-55 tubing always corrode preferentially to the J-55 grade and at fairly rapid rates. This appears to be a galvanic affect, although other experience shows that N-80 steel in the hole fails much faster than J-55 in wet CO_2 service. This may be the result of the quenched and tempered microstructure of N-80 steel, which is more susceptible to corrosion than the as-rolled structure of J-55.

As described before, ringworm corrosion is specific to upset J-55 tubing that has not been full length normalized. Figure 4–13 shows a typical example of ringworm corrosion caused by wet CO_2. This selective attack just beyond the upset is attributed to a galvanic effect between the different microstructure in the upset heat-affected zone and the nonaffected tubing body.

Corrosion of steel in the presence of carbon dioxide is a function of both chromium and nickel content as shown in Figure 4–14. At least 9% Cr is necessary to gain good corrosion resistance, but more often the threshold chromium content to ensure resistance is 12%. Thus 12–13% Cr steels represent the first step in applying metallurgy to tubulars for resistant to CO_2 corrosion. As was demonstrated in Figures 4–11a and 4–11b, the corrosion rate of steel in carbon dioxide is dramatically reduced with the addition of 9% Cr or greater to the steel. However, if the temperature is increased to 250°C (see Figure 4–11b), the 9 Cr steel corrodes at greater than 10 mm/y (395 mpy) while the 13% Cr corrodes much less and the 25 Cr-6 Ni (duplex stainless steel) corrosion rate remains almost zero. Thus increased resistance to CO_2 corrosion is a strong function of chromium content even at high temperatures and CO_2 pressure.

To this point, the discussion of alloying for resistance to CO_2 corrosion is quite straightforward; however, as the environment becomes more complex, so do the metallurgical solutions. If chlorides are included in the carbon dioxide environment, as they almost always are in petroleum production, pitting corro-

Figure 4–13 Ringworm corrosion adjacent to the upset in J–55 tubing.

sion and stress-corrosion cracking from chlorides becomes a concern for many alloys. Since stress-corrosion cracking typically initiates from pits on the surface of the alloy, pitting will be discussed in conjunction with SCC even though it must be kept in mind that failure can also occur strictly by pitting without associated SCC.

Figure 4–15 shows the regions of applicability of 13 Cr and 22 Cr stainless steels as a function of sodium chloride concentration and temperature. The 22 Cr is a duplex stainless steel (a combination of austenite and ferrite) with approximately 6% Ni. These alloys have a two-phase structure as connoted by the name *duplex*. The phases are ferrite and austenite and are approximately balanced with 50% ferrite and 50% austenite. Generally, with increasing chlorides and temperature the corrosion rate increases for 13 Cr but the 22 Cr shows much better resistance to corrosion. However, there is a penalty in changing from 13 Cr to 22 Cr. Duplex stainless steels (22 Cr and 25 Cr) cannot be heat-treated for strength but must be cold-worked. Above about 350°F duplex stainless steels can display unacceptable corrosion rates in high CO_2, high chloride wells. Increasing nickel content in these stainless steels can initially have a detrimental affect on SCC resistance. See Figure 2–8 which shows the minimum resistance to SCC for Fe-Cr-Ni alloys at approximately 8% Ni. Thus, in increasingly severe environments containing chlorides at high temperatures, more resistant alloys must be selected that, according to Figure 2–8, will be primarily high in nickel content (>40% Ni) or nickel-base alloys. Many of the high-nickel and nickel-base alloys are not heat treatable and therefore must be cold worked to obtain the desired mechanical properties. The group of alloys ranging from

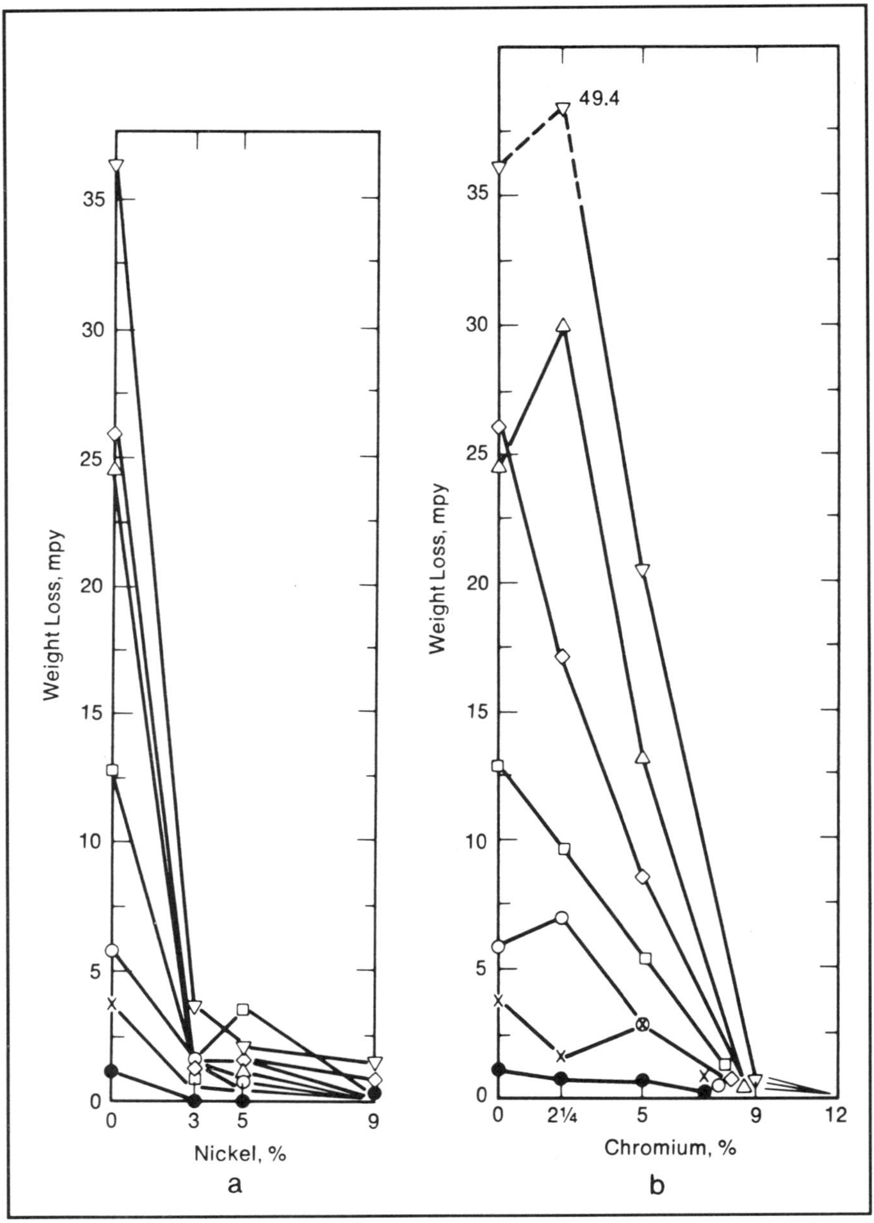

Figure 4–14 Effect of (a) nickel content and (b) chromium content on CO_2 corrosion of steels in gas condensate wells (After *NACE Pub. No. 50–3*).

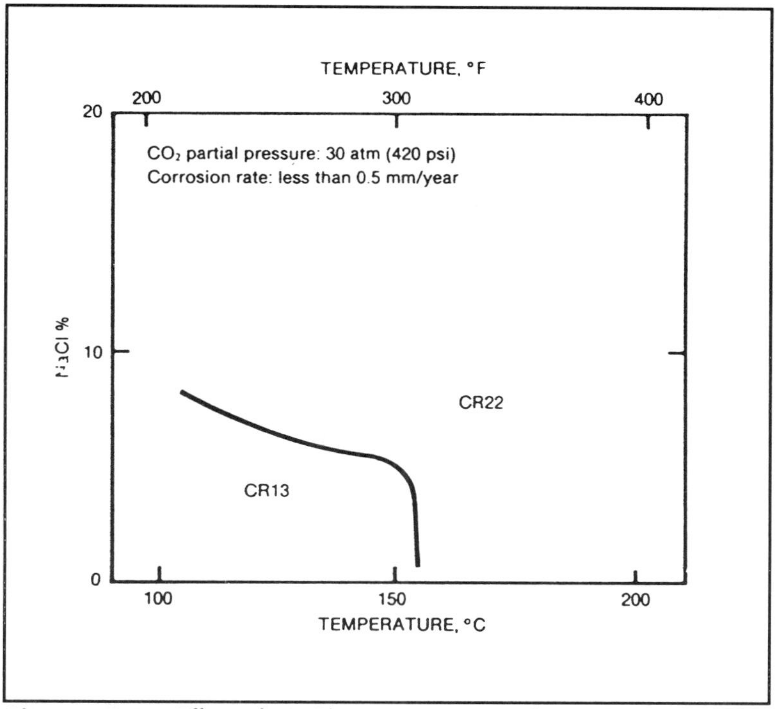

Figure 4–15 Effect of NaCl concentration and temperature on CO_2 corrosion of 13 Cr and 22 Cr stainless steels (After NKK).

stainless steels to high nickel content iron-base alloys, nickel-base alloys, cobalt- nickel alloys, and titanium alloys has become known as corrosion resistant alloys or CRAs. The chemical composition of some of the more common CRAs being used or considered for use in the industry are presented in Table 4–3.

The SCC behavior of some of these alloys in high-temperature brines is shown in Table 4–4. The alloys are listed in order of increasing alloy content. The strong effects of nickel above 30% and molybdenum in reducing SCC are evident. Therefore, as the bottom-hole temperature exceeds 350°F in the presence of chlorides, nickel-base alloys are needed to avoid chloride SCC.

The addition of hydrogen sulfide to a system containing carbon dioxide and chlorides creates a complex environment for which tubular selection becomes quite difficult. Before discussing this problem, it is beneficial to examine the effect of wet hydrogen sulfide alone on steels.

Hydrogen sulfide in oil and gas production can be detrimental in two ways: it is an aggressive pitting agent, and it induces cracking in many alloys. The general weight-loss corrosion resulting from hydrogen sulfide corrosion is con-

Table 4–3 Nominal Composition of Common CRAs

CRA Type	Common Name	UNS No.	Element (wt. %)								Strengthening Method
			Cr	Mo	Ni	Ti	Co	Fe	Cb	Other	
Stainless Steels	13 Cr (420)	S42000	13.0	...	0.5	...	...	Bal	...	...	Heat Treat
	316	S31600	17.0	2.5	9.0	...	...	Bal	...	...	Cold Work
	DP3 [1]	S31260	25.0	3.0	7.0	...	...	Bal	...	...	Cold Work
	Cr22 [2]	S31803	22.0	3.0	6.0	...	...	Bal	...	...	Cold Work
	AF22 [3]	S31803	22.0	3.0	6.0	...	...	Bal	...	...	Cold Work
	CD4MCu	None	25.0	2.0	5.0	...	...	Bal	...	3.0 Cu	Cold Work
	Nitronic 50 [4]	S20910	21.0	2.0	12.5	...	...	Bal	...	5.0 Mn	Cold Work
	Ferralium 255 [5]	S32550	25.0	3.0	5.5	...	...	Bal	...	1.0 Cu	Cold Work
	A286	S66286	15.0	1.2	25.0	2.5	...	Bal	...	...	Heat Treat
	17-4PH	S17400	17.0	...	4.0	...	...	Bal	...	4.0 Cu	Heat Treat
	Sanicro 28 [6]	N08028	27.0	3.5	31.0	...	...	Bal	...	1.0 Cu	Cold Work
	254 SMO [7]	S31254	20.0	6.1	18.0	...	...	Bal	...	0.2 N, .75 Cu	Cold Work
	20Cb-3 [11]	N08020	20.0	3.0	35.0	...	...	Bal	...	...	Cold Work
Nickel Alloys	NIC 42 [2]	N08825	22.0	2.9	42.0	...	...	Bal	...	...	Cold Work
	Incoloy 825 [8]	N08825	20.5	3.0	42.0	...	...	Bal	...	...	Cold Work
	Incoloy 925 [8]	N09925	21.0	3.0	42.0	2.1	...	Bal	...	2.0 Cu, 0.3 A1	Heat Treat
	All Corr [12]	N06110	31.0	10.0	56.0	...	...	Bal	...	2.0 W	Cold Work
	Inconel 718 [8]	N07718	19.0	3.1	52.0	0.9	0.5	Bal	5.0	0.6 A1	Heat Treat
	Inconel X-750 [8]	N07750	15.0	...	Bal	2.5	1.0	7.0	1.0	...	Heat Treat
	Inconel 625 [8]	N06625	22.0	9.0	Bal	0.4	1.0	5.0	4.0	...	Cold Work
	Monel 400 [8]	N04400	...	...	Bal	...	...	2.3	...	32.0 Cu	Cold Work
	Monel K-500 [8]	N05500	...	...	65.0	0.5	...	1.0	...	2.9 Cu	Heat Treat
	Pyromet 31 [11]	N07031	23.0	2.0	56.0	2.5	...	15.0	1.0	1.5 A1	Heat Treat
	SM 2550 [1]	N06975	25.0	6.2	49.0	1.2	...	...	...	...	Cold Work

	NIC 52 [2]	N06975	25.0	7.0	52.0	1.0	. . .	Bal	. . .	1.0 Cu	Cold Work
	Hastelloy C-276 [9]	N10276	15.4	16.0	Bal	. . .	2.0	6.0	. . .	3.5 W	Cold Work
	Hastelloy G-3 [9]	N06985	22.0	7.0	Bal	. . .	3.0	19.5	. . .	1.0 W	Cold Work
	Hastelloy G-50 [9]	None	22.0	9.0	Bal	. . .	3.0	18.0	. . .	2.0 W	Cold Work
Nickel/Cobalt	MP35N [10]	R30035	20.0	9.5	35.5	. . .	35.0	. . .	. . .	. . .	Cold Work
	MP 159 [10]	None	20.0	7.0	25.5	3.0	35.0	8.6	. . .	. . .	Cold Work
Cobalt Alloys	Haynes Alloy 25 [9]	R30605	20.0	. . .	10.0	. . .	Bal	3.0	. . .	15.0W	Cold Work
	Haynes Alloy 188 [9]	R30188	22.0	. . .	22.0	. . .	Bal	3.0	. . .	14.0 W	Cold Work
Titanium Alloys	Ti-6A1-4V-0.05 Pd	R56400	. . .	. . .	. . .	Bal	. . .	. . .	. . .	6 A1, 4 V, 0.05 Pd	Heat Treat
	Beta C [13]	R58640	6.0	3.0	. . .	Bal	. . .	. . .	. . .	3 A1, 8 V, 4 Zr	Heat Treat

1. Sumitomo Metal Industries, Ltd.
2. Nippon Kokan K.K.
3. Mannesmann
4. Armco Steel Corporation
5. Bonar Langley
6. Sandvik AB
7. Avesta Jernverks, AB
8. Huntington Alloys, Inc.
9. Haynes
10. Standard Press Steel
11. Carpenter Technology Corporation
12. Teledyne Allvac
13. RMI

Table 4–4 SCC of Alloys in Various Brines at Different Temperatures and 7-Day Exposure

	$CaCl_2$					NaCl Temperature °F (°C)					$MgCl_2$				
Alloys	**250 (121)**	**300 (149)**	**350 (177)**	**400 (204)**	**450 (232)**	**250 (121)**	**300 (149)**	**350 (177)**	**400 (204)**	**450 (232)**	**250 (121)**	**300 (149)**	**350 (177)**	**400 (204)**	**450 (232)**
Ferralium 255[1] (25.6–5.17–3.3)*				025	025				25	025			5	025	025
Sanicro 28[2] (26.4–31.4–3.0)			2	025	025				025	025			5	025	025
Incoloy 825[3] (21.4–42.0–2.5)								0		025				05	025
Hastelloy G-3[4] (22.5–43.3–7.1)										025					025
Hastelloy C-276[4] (15.2–54.8–16.1)															025

*Values are for alloy content in the order, chromium-nickel-molybdenum
Note: 0,2,3, and 5 indicate cracks in the annealed, 20%, 30%, and 50% cold-worked condition
[1]Bonar Langley Alloys Ltd.
[2]Sandvikens Jernverks Aktiebolag
[3]INCO International
[4]Haynes
(After Kolts)

siderably less than that caused by carbon dioxide at the same partial pressure for several reasons. A semiprotective film of iron sulfide (FeS) sometimes forms on steel during corrosion and H_2S dissolved in water does not form a simple acid as does CO_2 (carbonic acid). Hydrogen sulfide also enhances pitting attack of many alloys, especially steels.

One of the greatest concerns facing the industry, however, is catastrophic cracking in hydrogen sulfide environments. This phenomenon has been determined to be a special case of hydrogen-stress cracking that is more severe than for many other hydrogen-bearing environments because hydrogen sulfide poisons the hydrogen recombination reaction occurring at the cathode, thereby charging atomic hydrogen into the alloy. Sulfide-stress cracking (SSC) follows the same dependency of parameters as hydrogen-stress cracking. Thus, it has a temperature-versus-time-to-failure minimum at about room temperature. The significance of this to oil and gas wells is that susceptibility to SSC is greatest at the top of a casing or tubing string and on the surface equipment. As the applied stress and yield strength increase, the susceptibility to SSC increases.

The effect of temperature on SSC of tubulars is shown in Figure 4–16. This figure also shows the effect of normalized and normalized and tempered J-55, C-75, and N-80 steel compared to quenched and tempered steels, reconfirming the effect of metallurgical structure on resistance. Furthermore, as the strength level increases, the temperature at which SSC does not occur increases. The practical benefit of this is that higher strength casing can be used deeper in a well where temperatures are higher.

Figure 4–17 shows the effect of increasing hydrogen sulfide concentration on SSC of common tubular grades. Again, the behavior of normalized and normalized and tempered steels is inferior to the quenched and tempered steels. Also, the high-strength steels tolerate less hydrogen sulfide before cracking. An example of a field failure of L-80 tubing by SSC is presented in Figure 4–18. Failure was principally the result of the deep slip marks shown on the OD of the pipe. Even highly resistant grades of tubing and casing can fail if improperly handled. Figure 4–19 shows the effect of notch depth and wall thickness on the threshold stress for SSC. The two modified N-80 grades are similar to L-80 while the other is a C-90 grade. Increasing notch depth with increasing wall thickness can reduce the threshold stress for SSC more than 50%; for example, compare a 5% notch in 0.250-in. material with a 12 1/2% notch in the same material at a thickness of 0.850 in.

The composition of steel is also quite important in determining resistance to SSC. Figure 4-20 compares the behavior of 4130 low-alloy steel to that for C-Mn steels. The former steel has better resistance to SSC because of its greater hardenability and its resistance to tempering. This sluggish tempering behavior allows the 4100 series alloys to be tempered at higher temperatures than the

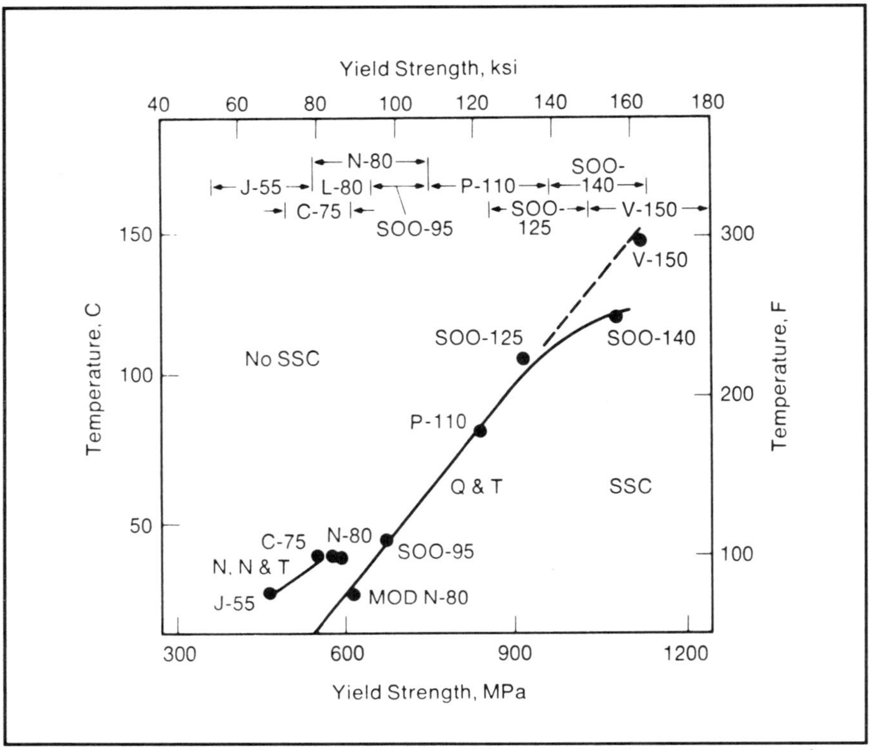

Figure 4–16 Effect of temperature on sulfide-stress cracking resistance of common grades of tubing and casing (After Kane and Greer, © 1977, SPE).

carbon-manganese steels to obtain the same hardness and mechanical properties. The higher-temperature tempering allows for more recovery of the microstructure and thus a uniform structure, which provides greater resistance to SSC. The relationship of hardenability to SSC resistance is shown in Figure 4–21. The ability to achieve complete transformation to martensite is a function of hardenability of the alloy and the speed of the quench. The highest quality steels for SSC resistance have been quenched to near 100% martensite. The benefits of achieving near 100% martensite during the quenching are so great, API 5CT requires a hardness test be performed and a minimum hardness on as-quenched C–90 grade tubulars be attained in compliance with the following formula: $HRC_{min} = 50 \times \%C + 29$. This minimum hardness represents a minimum of 90% martensite. (Note that carbon content of the steel is the sole factor in determining the as-quenched hardness.)

A generally accepted rule for resistance of carbon and low-alloy steels to SSC is to maintain the hardness below HRC 22. While there are exceptions to this

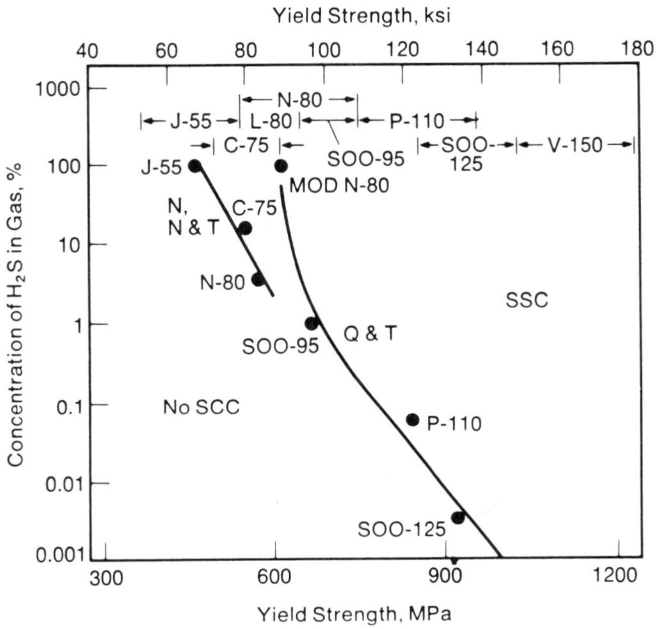

Figure 4–17 Effect of H_2S concentration on sulfide-stress cracking of tubing and casing steels (After Kane and Greer, © 1977, SPE).

rule, it has been very effective over the years and is satisfactory if no data or experience exists to contradict it for a specific application. Steels based on the 4100 series are acceptable up to HRC 26 provided they are tubulars. Higher alloyed materials can be exposed to sour (H_2S containing) environments at higher hardnesses. The acceptable hardness limits for many alloys to reduce the risk of SSC are described in *NACE Specification MR-01-75.* This specification is frequently misapplied to include weight loss corrosion from H_2S and other forms of environmental cracking such as SCC. MR-01-75 is only applicable to the SSC resistance of alloys.

These upper limits on hardness necessitate a maximum on the tensile strength and, therefore, yield strength of an alloy that can be used in hydrogen sulfide service. One means around the strength limitation is to increase the wall thickness of a tubular. Since stress is the force or load applied over a certain area (in this case, the cross-sectional area of the tube wall), increasing the wall thickness of the tubular can substantially reduce the applied stress, allowing

Figure 4–18 Sulfide-stress cracking failure of L-80-grade tubing. Note deep slip marks that assisted fracture initiation.

lower-strength tubulars to be used. It must be kept in mind that increasing the wall thickness also increases the weight of the joint, thereby also increasing the applied load on top of the string. These two opposing factors must be balanced to arrive at an optimum string design for reducing the potential for SSC. The weight of the string can be partly offset by using a tapered string where possible (i.e., using successively smaller pipe toward the bottom of the string). It is

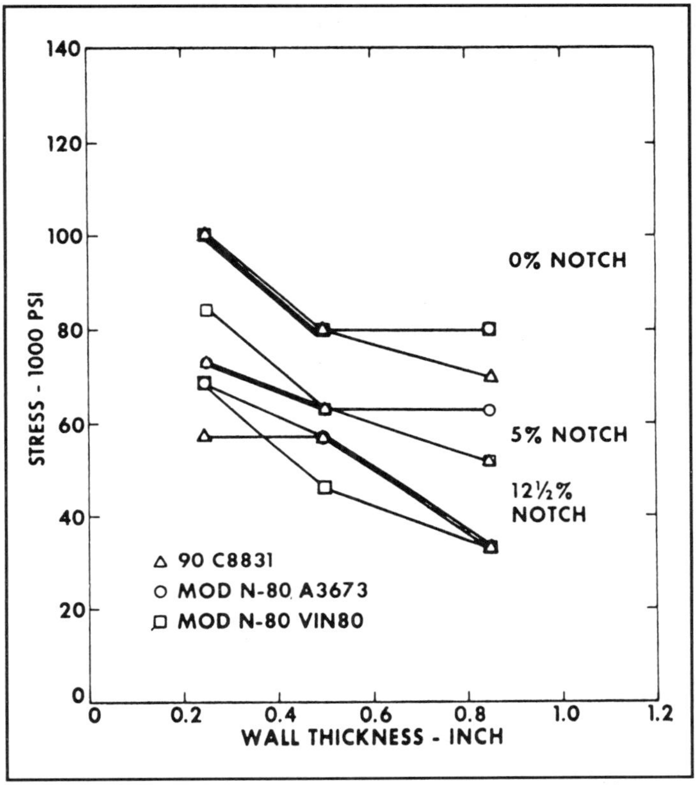

Figure 4–19 Effect of wall thickness and Charpy notch depth on threshold stress for sulfide-stress cracking (After Greer, © SPE).

always recommended to use the lowest-strength tubular that can be accommodated when designing and running strings in sour service.

The increase in deep drilling has presented a terrific challenge to completion and production of these wells. Since the top portion of the string may be required to support 5–6 miles (8–10 km) of pipe, the yield strength of the material must be high or the wall thickness quite large. High yield strengths are more susceptible to SSC and require less hydrogen sulfide to promote SSC, therefore, heavy walls are often the choice. This presents the problem of producing tubulars of L-80 through T-95 in heavy-wall sections, with a uniform microstructure and hardness across the wall.

Heavy-wall, low-alloy tubulars and couplings are often produced from 0.750–1.50 in. wall thickness. To achieve the desired properties in this kind of section thickness, the hardenability of the alloy becomes important. As discussed in

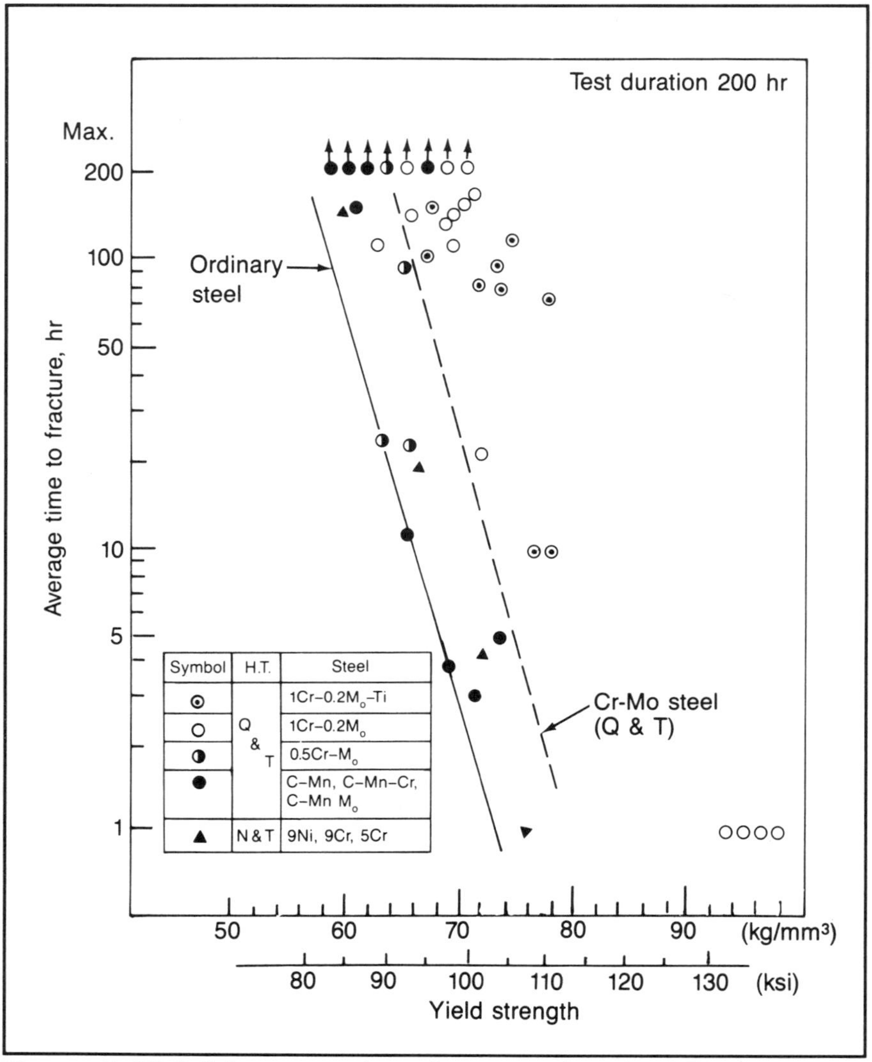

Figure 4–20 Sulfide-stress cracking of commercial grades C-75, N-80, L-80, C-95 (After Ikeda et al.).

Chapter One, the first step is a more rapid quench. This entails a water quench that, in some mills, is applied just to the external surface of the tubes. In other mills, both external and internal water quenching are performed. Looking at the end-quench data for 1040 steel (e.g., J-55) shown in Figure 4–22, the depth at which an external water quench will produce significant hardening is less than 3⁄16 in. Above and beyond the as-quenched hardness, there must also be allow-

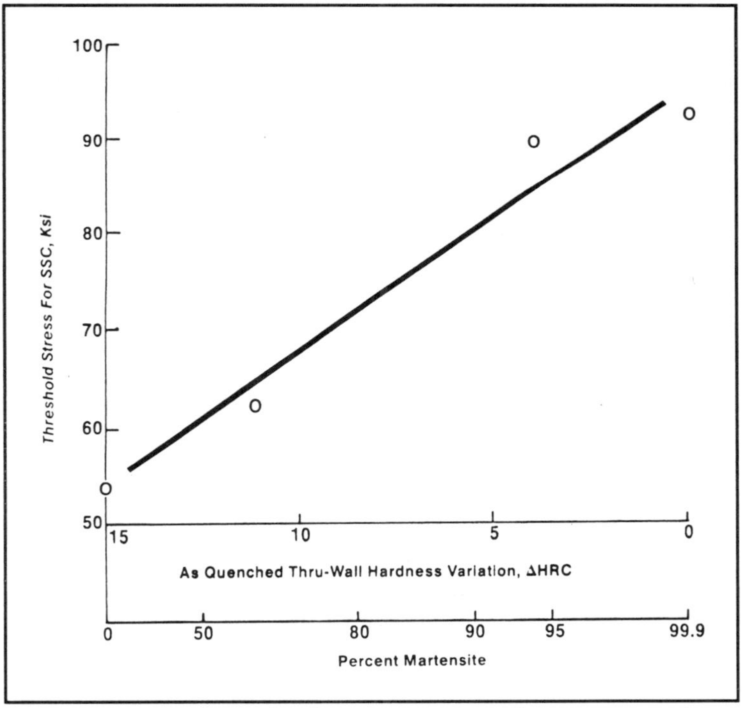

Figure 4–21 Relationship between percent of martensite and sulfide-stress cracking threshold for an AISI 4130 bar heat-treated to Rockwell C25 hardness (After Greer and Holland, © 1980, SPE).

ance for loss in hardness caused by tempering. Therefore, to obtain a fully martensitic structure and to allow sufficient hardenability such that tempering will not result in a product below needed strength levels, an alloy steel is required. Most tubulars that are used with large wall thickness and/or that require the desired SSC resistance are based on the AISI 4130 grade low-alloy steel. To compare the effect of alloying while keeping carbon constant, Figure 1–28 shows an end-quench hardenability curve for AISI 4140H in which the carbon content is kept constant. Compared to Figure 4–22, the hardness at 3/16 in. in the middle of the band for 1040 is HRC 33, whereas for 4130 the depth at which HRC 32 is achieved is about 1 7/8 in. Also, if the quench is inside/outside, the effective thickness to obtain HRC 32 is doubled. Thus the power of alloying elements in enhancing hardenability is tremendous. Those mills that have only an external quench capability often increase the carbon and manganese content or use alloying elements such as chromium, boron, or titanium to increase the hardenability.

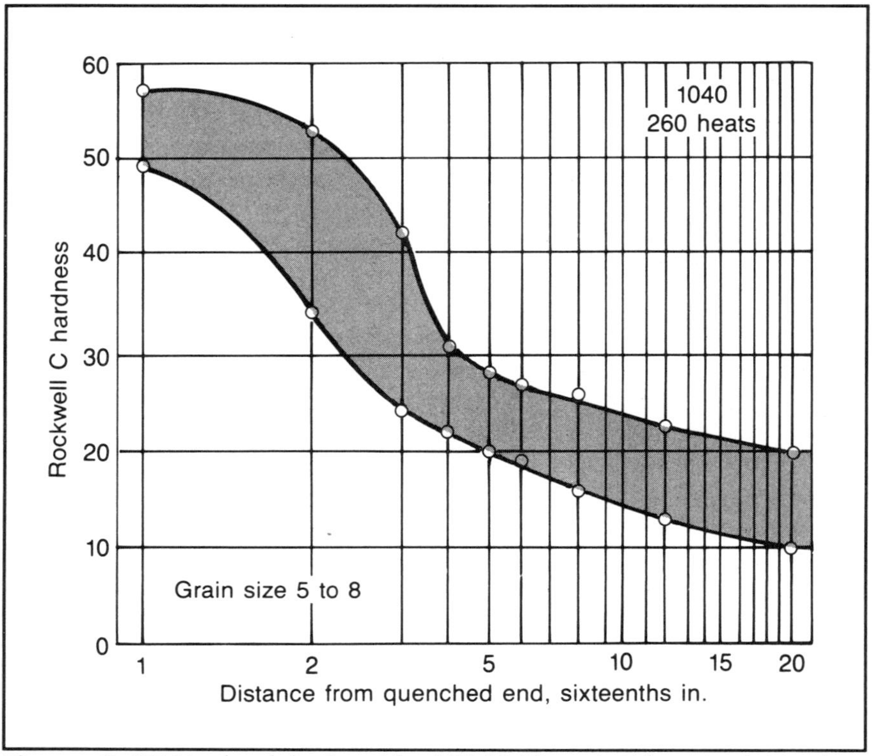

Figure 4–22 Hardenability band for AISI 1040 steels (After *Metals Handbook*).

In spite of attempts to push the usable strength of steel tubulars to higher limits in hydrogen sulfide, the threshold remains around 100 ksi. In fact, all low-alloy steels heat-treated to yield strengths in excess of about 135 ksi fail rapidly when exposed to hydrogen sulfide. Based on the earlier discussion of increasing resistance to corrosion and SSC with increased alloying, it would be expected that 13% Cr would have greater resistance to SSC than a low-alloy steel. Figure 4–23 shows this assumption to be incorrect, and that AISI 410 (12 Cr) or AISI 420 (13 Cr) stainless steel actually is more susceptible to SSC than the low-alloy steels. This is usually attributed to the difficulty in producing a completely tempered martensitic structure. Quenching followed by a single temper often leaves a significant amount of untempered martensite that can cause a reduction in resistance to SSC. Thus, a double temper is required with the second temper specified at 50°F less than the first. Still, the resistance of 13 Cr to environments containing H_2S is marginal, especially in the presence of chlorides.

The duplex stainless steels offer better chloride-stress cracking resistance

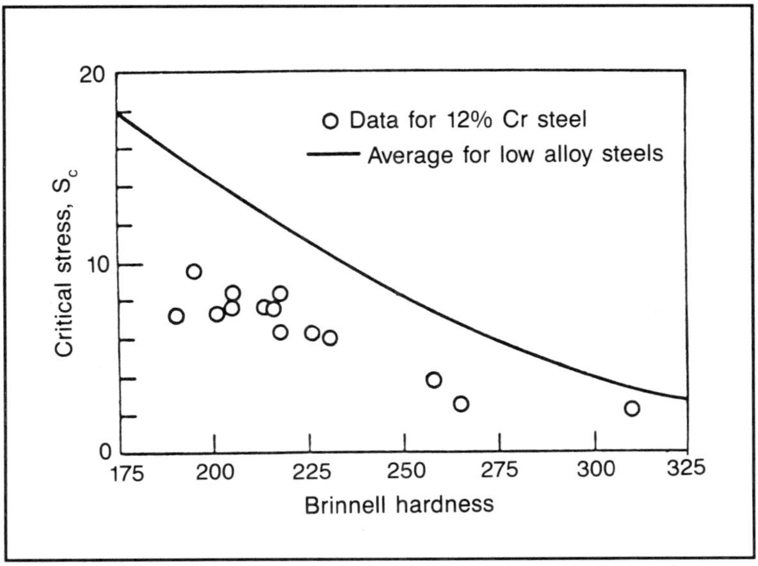

Figure 4–23 Comparison of the sulfide cracking tendency for 12% chromium versus low-alloy steel. S_c is a measure of resistance to SSC. Increasing S_c is equivalent to increasing resistance to SSC (After Treseder and Swanson).

than austenitic stainless steels and enhanced weight-loss corrosion resistance from CO_2 than 13% Cr steel. They also show increased SSC resistance; however, this resistance to SSC is limited to low concentrations of hydrogen sulfide on the order of 1.0 psia or less. Of course, this depends on the amount of cold work as well as other factors. As the hydrogen sulfide content increases, the suitability of duplex stainless steels diminishes (see Fig. 4–24), so high-nickel alloys must be utilized—especially when carbon dioxide and chlorides are also present in the environment.

Another consideration when using duplex stainless steels is that they are most often used in the cold-worked condition. Moreover, when subjected to high well temperatures, a certain amount of recovery occurs, leading to loss in strength (see Fig. 4–25). This loss in strength is common for many of the cold-worked CRA materials.

The behavior of CRAs in complex well environments has been extensively studied, and CRA tubulars have been run in many wells.

Figure 4–26 shows the threshold stress for SSC of some of the more common CRAs as a function of yield strength. These alloys were tested in the NACE solution (oxygen-free water containing 5% sodium chloride, 0.5% acetic acid,

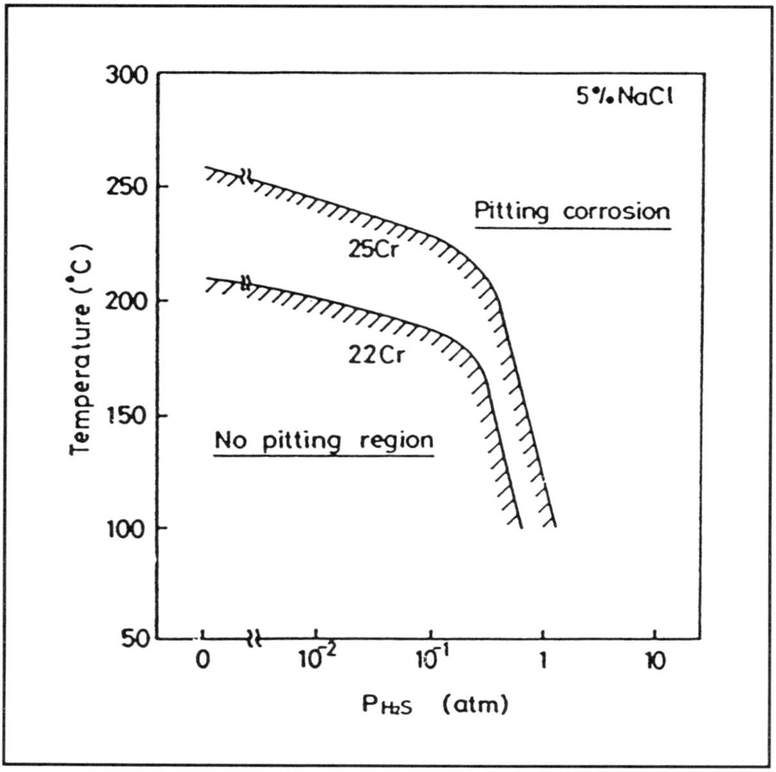

Figure 4–24 Influence of partial pressure of H_2S and temperature on pitting behavior of duplex stainless steels (After Sumitomo).

and saturated with hydrogen sulfide) at room temperature. None of these alloys failed in 30 days of exposure when stressed to 98% of their yield strength. Based on the information shown in Figure 4–26, it would appear that MP35N and the other alloys are essentially immune to failure in hydrogen sulfide. However, these materials are not totally immune to failure in aggressive well environments.

In fact, at temperatures above ambient up to bottomhole temperature, which currently can approach 500°F, cracking and corrosion of CRAs are necessary considerations. It has been found that cracking of CRAs at elevated temperatures in CO_2, H_2S, chlorides, and free sulfur cannot simply be distinguished as SCC or SSC. Therefore, cracking of CRAs in aggressive wells or surface equipment is referred to as environmental cracking (EC).

The selection of a suitable CRA tubular for a specific well environment depends on many factors such as gas composition, water chemistry, pressure,

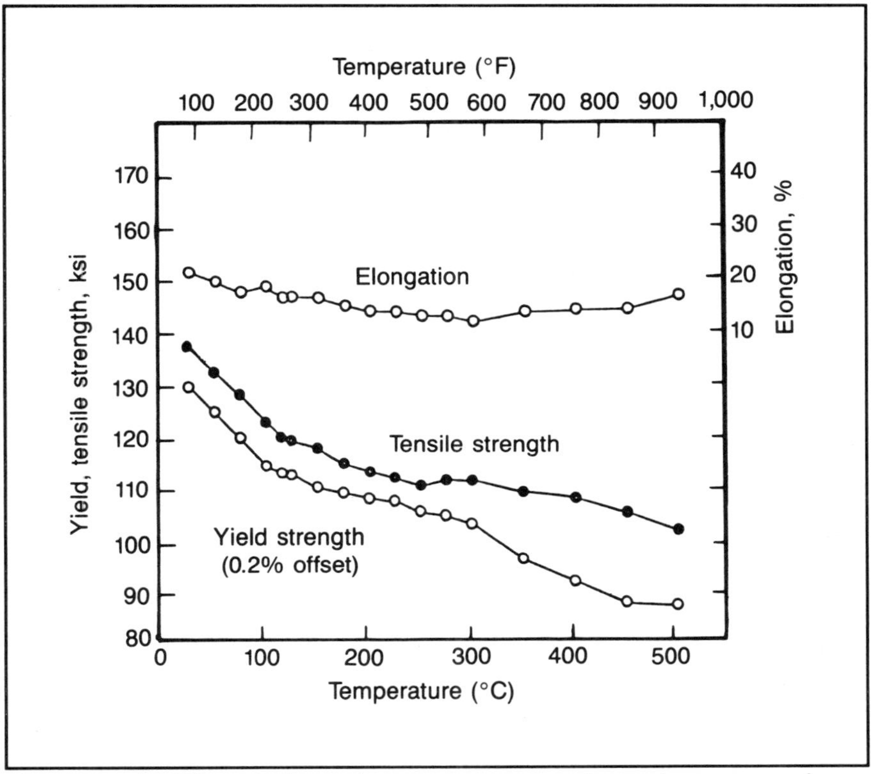

Figure 4–25 Effect of elevated temperature on mechanical properties of duplex stainless steel (After NKK).

temperature, risk, economics, etc. These factors are dealt with in more detail elsewhere (see for example, B.D. Craig, *SPE Monograph*).

Generally, the ranking of CRAs according to corrosion resistance is as follows in increasing order of resistance: martensitic stainless steels (13% Cr), duplex stainless steels, high nickel content iron-base alloys (e.g., Alloy 28, Alloy 825), nickel-base alloys with low molybdenum (e.g., Alloy 2550, G-3), nickel-base alloys with high molybdenum (Hastelloy C-276), nickel-cobalt alloy (e.g., MP35N) and titanium alloys. The last three alloy groups are generally equivalent although certain Ti alloys may have less resistance to the wellbore environment than others. Figure 4–27 shows the environmental cracking resistance of several titanium alloys under simulated sour well conditions.

The advantage of some titanium alloys over many nickel-base alloys is that they can be heat-treated to very high strength levels (>150,000 psi yield) so that thinner wall tubulars can be run at a lower cost than a comparable thick wall

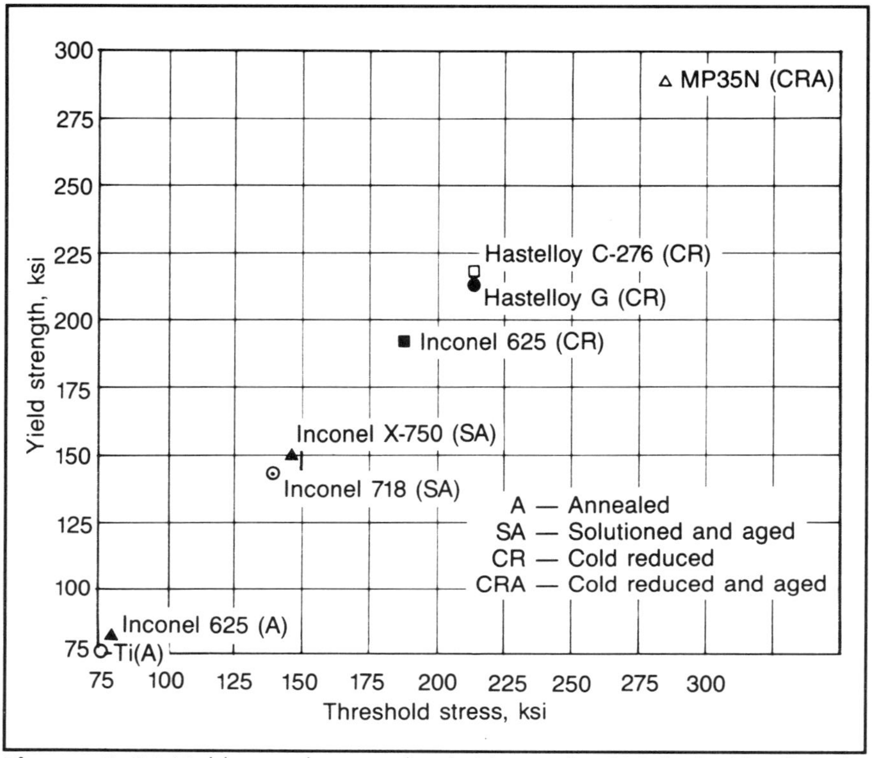

Figure 4–26 Yield strength versus threshold stress for SSC for highly alloyed materials. None of the alloys failed in the NACE test (data from Watkins and Greer).

nickel alloy. However, the acid resistance of titanium alloys is poor compared to the nickel alloys, and their fracture toughness is considerably less than for nickel base alloys.

If not properly manufactured, stainless steels and nickel alloys may contain a variety of phases that are detrimental to the performance of these alloys. Some of these phases are carbides ($M_{23}C_6$, M_7C_3, M_6C, MC, where M = Cr, Fe, Mo, Ti, Nb, etc.) and nitrides (Cr_2N) or compounds such as sigma, mu and Laves phases. These latter compounds can have a deleterious effect on the fracture toughness as well as the corrosion resistance of the stainless steel and nickel alloys in which they form.

Sensitization as discussed in Chapter Two can occur in many of these alloys and reduce their corrosion resistance. Figure 4–28 is an example of solution-annealed Hastelloy C-276 tubing prior to final pilgering for size and strength. Figure 4–29 shows a similar nickel-base alloy in the same condition that reveals

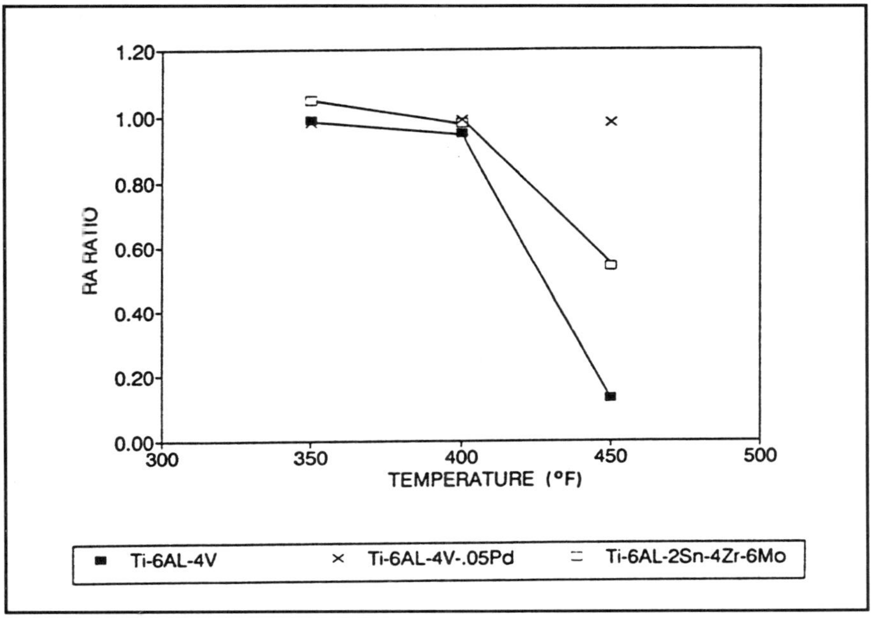

Figure 4–27 Environmental cracking of three Ti alloys measured as reduction in area ratio versus temperature (After Sumitomo).

significant grain boundary carbide precipitation. The latter microstructure is far less suitable for exposure to a corrosive environment than the C-276 in Figure 4-28.

Figure 4–30 shows the same alloy in Figure 4–28 after pilgering to achieve the desired strength. Note the elongated grains and the small parallel lines (twins) that are indicative of a cold-worked alloy.

Hydrogen-stress cracking is a potential problem in service for alloy tubulars since many of the other components in a system may be carbon steel (i.e., casing, hangers, etc.). Coupling between these alloys and steel may lead to hydrogen charging of the CRA. Nickel alloys show greatest susceptibility to HSC around room temperature. An example of the effect of high-temperature aging on HSC resistance of a nickel alloy is shown in Figure 4–31.

Stress-corrosion cracking (SCC) can also become a problem for CRAs as the temperature increases. This aspect was discussed earlier in Table 4–4 which demonstrates the effect of cold work and temperature on SCC. Table 4–5 substantiates this SCC problem with data from tests in a much more severe environment. The effects of increasing temperature and applied stress are evident. Although these data indicate somewhat poor resistance of nickel- and cobalt-base alloys to HSC and SCC, it must be kept in mind that these are laboratory

Table 4–5 Stress-Corrosion Cracking of Nickel- and Cobalt-Base Alloys (C-Ring Specimens Tested in Aqueous Solutions Containing 25% NaCl, 0.5% Acetic Acid, 1 g/l Sulfur, and H_2S)

		Test Temperature, °C (°F)					
		175 (350)		205 (400)		290 (550)	
Material	**% Trans. Y.S.**	**70**	**90**	**70**	**90**	**70**	**90**
Hastelloy Alloy C-276 (37% CR)		0	X	X	X	X	X
MP35N (59% CR)		X	X	X	X	X	X
Hastelloy Alloy G (59% CR)		...	...	X	X	...	...

0 = Passed
X = Failed
Time to failure by SCC was 100 days in some cases. Test duration needed to obtain data shown in this table was 150 days.
(After Kane and Boyd)

tests and therefore intentionally more severe than actual conditions. They are meant to examine parameters that affect cracking behavior in a relatively short testing time. The fact is that many of the alloys discussed have been used successfully in field applications. What is important to gather from this discussion of CRAs is that many factors affect corrosion and cracking resistance, and when an alloy must be selected for a particular application, all of these factors must be considered.

The advent of CRA tubulars has brought with it another problem not common to normal oilfield tubulars—galling. Stainless steels are notorious for thread galling and, as expected, tubulars made of these alloys suffer from the same problem. Several field applications of 13% Cr tubulars have been plagued with galling of the joints, as have the duplex stainless steels. Since galling is a function of surface preparation, work has been done recently on enhancing antigalling by special surface preparation, including the application of copper plating on the pin threads.

One alternative to solid CRA components is the use of CRA clad over carbon and low-alloy steel. While these materials are available, the industry has been slow to adopt them, especially for downhole applications. Clad tubing has been run and performed successfully in a few instances. The obvious cost advantage over solid CRA should provide an impetus for further use in the future.

From the different possible corrosion-related failure mechanisms discussed,

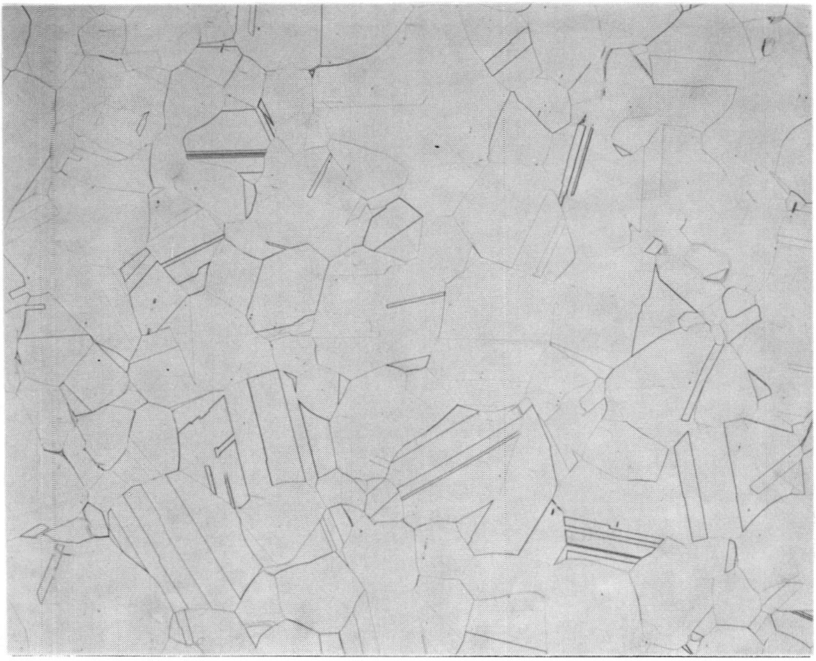

Figure 4–28 Microstructure of solution-annealed Hastelloy C-276 showing a single phase with essentially no carbide precipitation. Mag. 100X.

it is evident that the design of a tubular string must take into account all of these factors.

At the top of the hole where the string may be close to ambient temperature, failure by SSC is most probable, while at the bottom of the hole where the temperature is greatest, chloride cracking is of primary concern. Between these two extremes, failure may occur by either or both mechanisms. Pitting attack and general weight-loss corrosion will be greatest near the bottom of the string because of temperature. But since water may not drop out as a liquid until farther up the string, these problems can be expected almost anywhere in the well.

One can appreciate from this discussion of tubulars that no set rules exist for materials selection for tubing and casing. Furthermore, many corrosion mechanisms may be operative at the same time, so each must be considered and the likelihood of its contribution weighed against the cost of alloying to avoid failure. Additionally, the demand for CRA tubulars is often beyond the capability of many mills to produce in the sizes, weights, and quantities desired of a certain alloy; thus, availability can be a problem. Finally, but in some ways most important,

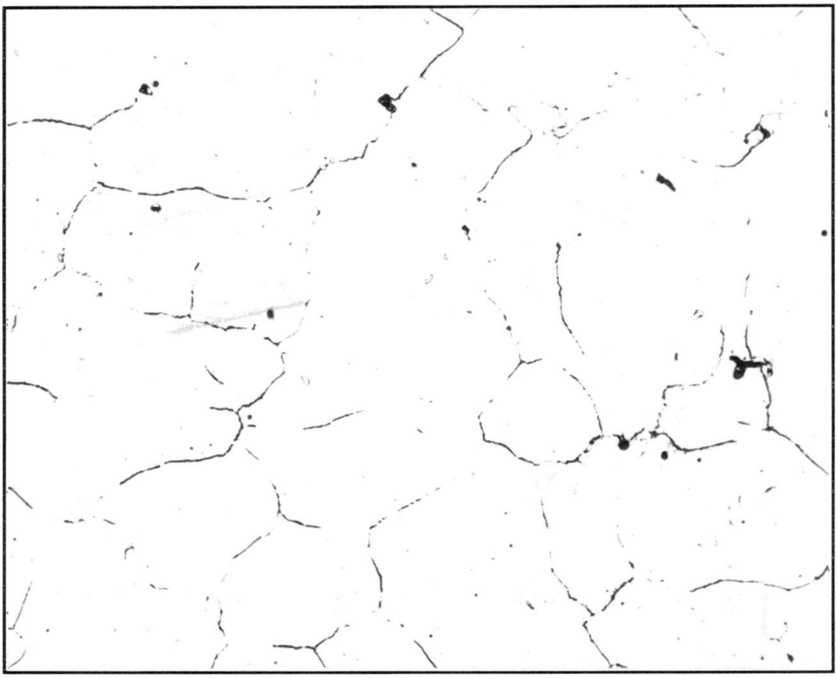

Figure 4–29 Microstructure of a nickel-base alloy similar to C-276 showing severe carbide precipitation along the grain boundaries. Mag. 100X.

the quality control of low-alloy steels and CRAs is crucial to good performance.

Sucker Rods

Sucker rods represent another kind of downhole equipment that frequently fails. The movement of sucker rods results in a cyclic load on the whole string and each rod individually as a result of the up-and-down stroke during pumping. In one complete cycle of the rod string, the rods go through several stress cycles because of the acceleration and deceleration of the string and the elastic elongation of each individual rod. The maximum stress occurs on the upstroke and the minimum stress on the downstroke. The loading on the entire string and on each rod individually is fatigue.

Sucker-rod metallurgy and properties are covered under *API Specification 11B* and the polished rod under *11D*. The chemical and mechanical properties are reproduced from the API in Table 4–6.

Table 4–6 Chemical and Mechanical Properties of Sucker Rods

Grade	Chemical Composition	Tensile Strength, psi Min.	Max.
K	AISI 46XX	85,000	115,000
C	AISI 1036*	90,000	115,000
D	Carbon or alloy**	115,000	140,000

* Not restricted to AISI 1036 steel
**Any composition that can be effectively heat treated to the minimum ultimate tensile strength
(After API 11B, table 2–1)

A compilation of mechanical and chemical properties of some of the commercially available sucker rods is shown in Table 4–7. Additional processing is included for some rods to increase the fatigue performance. This may be done by means of induction-hardening the surface or by shot-peening the surface. These two treatments place the surface in compression—in some cases as high as 125,000 psi for the induction hardening—which produces a case hardness of BHN 475–650. Compressive stresses of 67,000 psi have been measured after shot peening.

Figure 4–30 Micrograph of the same alloy as shown in Figure 4–28 after cold working to increase strength. Mag. 100X.

Table 4-7 Chemical Composition and Mechanical Properties of Common Commercial Sucker Rods.

Manufacturer	Designation	Alloy	C	Mn	P (Maximum)	S (Maximum)	Si	Ni	Cr	Mo	Other
			Chemistry (wt%)								
API GRADE C (carbon steel)											
Axelson	S-60	1036M	0.25-0.31	1.00-1.35	0.040	0.040	0.15-0.30				
LTV	1	1536	0.30-0.37	1.20-1.50	0.040	0.040	0.15-0.30				
Corod	CD	1036M	(0.34)	(1.33)	(0.18)	(0.18)	(0.26)				
Upco	35	1536	0.30-0.37	1.20-1.50	0.040	0.040	0.15-0.30				
Trico	C11	1536	0.30-0.37	1.20-1.50	0.040	0.040	0.15-0.30				
Norris	30	1536M	0.34-0.39	1.15-1.45	0.040	0.040	0.20-0.30				
Oilwell	C	1536	0.30-0.37	1.20-1.50	0.040	0.040	0.15-0.30				
VEW-SBS	VM125	1036	0.28-0.35	1.30-1.50	0.030	0.030	0.20-0.30				
API GRADE K (nickel-molybdenum alloy)											
Axelson	S-59	46XX	0.14-0.22	0.45-0.65	0.040	0.040	0.15-0.35	1.65-2.00		0.20-0.30	
LTV	5	4623	0.20-0.25	0.75-1.00	0.040	0.040	0.20-0.35	1.60-2.00		0.20-0.30	
Upco	45	4621M	0.20-0.25	0.60-0.80	0.025	0.035	0.20-0.30	1.60-2.00		0.15-0.25	
Norris	40	4621M	0.20-0.25	0.60-0.80	0.025	0.035	0.20-0.30	1.65-2.00		0.15-0.25	
VEW-SBS	N2P	4621	0.13-0.23	0.60-0.90	0.030	0.030	0.20-0.35	1.65-2.00	0.25 min	0.20-0.30	
API Grade D											
Carbon steel											
Axelson	X-67	1036M	0.25-0.30	1.00-1.35	0.040	0.040	0.25-0.30				
Upco	75	1541	0.36-0.44	1.35-1.55	0.030	0.030	0.15-0.30				0.07-0.10V
Chrome-molybdenum alloy											
LTV	3	4142M	0.39-0.46	0.65-1.10	0.040	0.040	0.20-0.35		0.75-1.20	0.20-0.30	
Upco	75	4142	0.39-0.46	0.65-1.10	0.040	0.040	0.20-0.35		0.75-1.20	0.15-0.25	
Trico	D-61	4142M	0.40-0.45	0.75-1.00	0.035	0.040	0.20-0.30		0.80-1.10	0.08-0.12	
Norris	78	4142M	0.40-0.45	0.80-1.00	0.035	0.035	0.20-0.30	0.45 max	0.90-1.00	0.15-0.25	
VEW-SBS	S200h	4135	0.30-0.37	0.50-0.70	0.030	0.030	0.15-0.25		0.90-1.20	0.25 max.	0.02-0.03Cb
Special alloy											
Axelson	S-87	Special	0.22-0.29	0.70-1.00	0.040	0.040	0.15-0.35	0.70-1.00	0.41-0.65		
Norris	90	Special	0.18-0.25	0.60-0.80	0.025	0.025	0.20-0.30	1.50-1.60	0.70-0.90	0.20-0.30	0.04-0.05V
Trico	D-63	4720	0.17-0.23	0.50-0.70	0.030	0.030	0.20-0.35	0.90-1.20	0.35-0.55	0.15-0.25	
MISCELLANEOUS											
Oilwell	EL	Special	0.31-0.38	0.60-0.90	0.040	0.040	0.15-0.30	1.25-1.75	0.60-0.90		
VEW-SBS	ARD3	DIN1.4122	0.35-0.42	0.50-0.80			0.30-0.50	0.70-0.90	16.00-17.00	0.90-1.10	

NOTES: M-modified, parentheses indicate typical values

Table 4–7 (continued)

Manufacturer	Designation	Yield Strength, 1,000 psi	Tensile Strength, 1,000 psi	Elongation 8 in., %	Reduction of Area, %	Brinell Hardness	Heat Treatment
API GRADE C (carbon steel)							
Axelson	S-60	90/105	100/115	13-20	65-71	207-235	Quenched & tempered
LTV	1	60/75	90/105	18-25	55-66	187-217	Normalized
Corod	CD	100 Min	115 Max		50-60	250-260	Quenched & tempered
Upco	35	60 Min	90/115	18-23	50-60	190-205	Normalized
Trico	C11	60/75	90/115	18-23	50-65	190-205	Normalized
Norris	30	60/75	90/110	18-23	50-65	190-205	Normalized
Oilwell	N	60/75	90/105	18-25	53 Min	185-217	Normalized
VEW-SBS	VM125	60 Min	90/115	18 Min	50 Min	185-215	Normalized
API GRADE K (nickel molybdenum alloy)							
Axelson	S-59	90/105	100/115	13-20	64-71	207-235	Quenched & tempered
LTV	5	70/85	88/105	16-25	60-70	182-217	Normalized & tempered
Upco	45	68/80	85/100	18-25	60-70	175-207	Normalized & tempered
Norris	40	70/80	90/100	18-25	60-70	175-207	Normalized & tempered
VEW-SBS	N2P	70 Min	85/115	18 Min	60 Min	175-205	Normalized & tempered
API GRADE D							
Carbon steel							
Axelson	S-67	110/125	120/135	11-17	55-63	248-277	Quenched & tempered
Upco	75	90/110	120/140	15 Min	45 Min	241-280	Normalized
Chromium-molybdenum alloy							
LTV	3	95/110	115/135	10-13	50-60	235-270	Normalized & tempered
Upco	75	100/115	120/140	10-15	50-65	241-280	Normalized & tempered
Trico	D-61	95/110	115/140	10-15	50-65	241-280	Normalized & tempered
Norris	78	100/110	120/140	13-18	50-60	241-280	Normalized & tempered
VEW-SBS	S200h	100 Min	115/140	12 Min	55 Min	235-285	Normalized & tempered
Special Alloy							
Axelson	S-87	110/125	120/135	12-17	57-67	248-277	Quenched & tempered
Norris	90	90/100	115/125	14-18	50-60	235-250	Normalized & tempered
Trico	D-63	95 Min	115 Min	14 Min	50-60	227-260	Normalized & tempered
MISCELLANEOUS							
Oilwell*	EL						Induction case hardened
VEW/SBS	ARD3	85 Min	115/135	14 Min	40 Min	235-275	Quenched & tempered

*General properties not applicable due to differences in case and core properties.

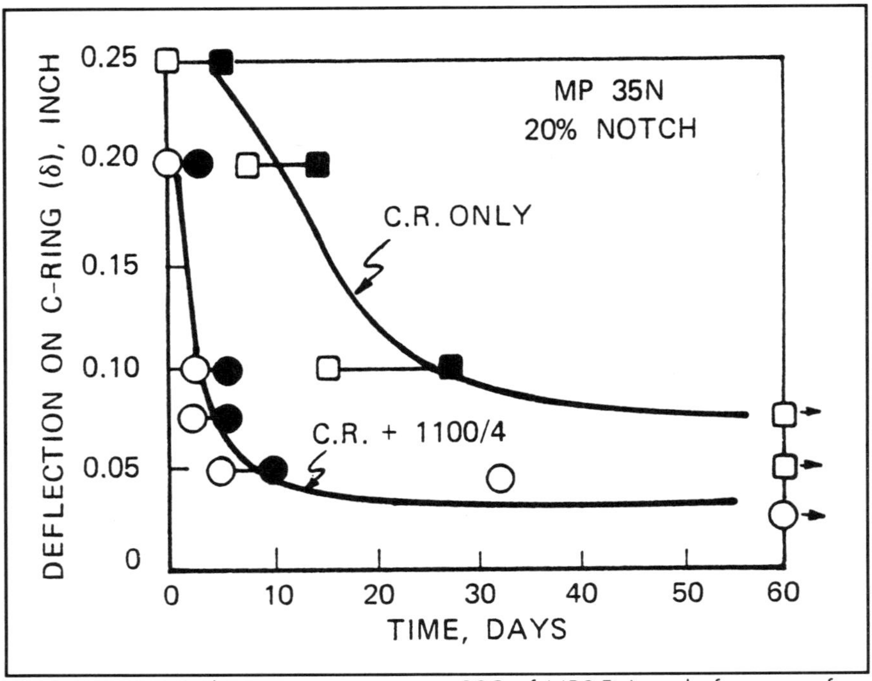

Figure 4–31 Reduction in resistance to SSC of MP35N steel after aging for 4 hr at 1,100°F. C-rings with carbon steel bolts subjected to room-temperature NACE solution (After Kane).

Couplings and subcouplings for pin-pin rods have essentially no chemical composition requirements and limited mechanical property requirements under the API. Class T couplings must be manufactured to a hardness range of Rockwell C 16–23. The sulfur content must not exceed 0.050%, which is the only chemical composition requirement.

Couplings are available in quenched and tempered AISI 4640 steels, which is case hardened to HRC 50–60 for wear resistance or spray metal coated with nickel-base powders (containing chromium, boron, and silicon) to a minimum thickness of 0.010 in. with hardnesses exceeding HV 595.

Polished rods used at the top of the string and run through the packing are manufactured from 1045 piston steel, ground and polished; 4140 low-alloy steel; cold-drawn bronze; cold-drawn Type 304 stainless steel; or Monel. They are also supplied with a hard facing or are chrome-plated.

Depending on the source of information, from 60–80% of all sucker-rod failures occur in the body; the remainder occur in the connection. These failures are predominantly the result of corrosion fatigue, as would be expected from the cyclic nature of their application. Figures 4–32 through 4–34 show examples of sucker-rod corrosion that resulted in corrosion-fatigue failures.

Occasionally, failure is the result of tensile overload and will display the necked-down area in the region of fracture. This is, of course, caused by exceeding the tensile strength of the rod and is commonly the result of pulling a subsurface pump that is stuck in the tubing.

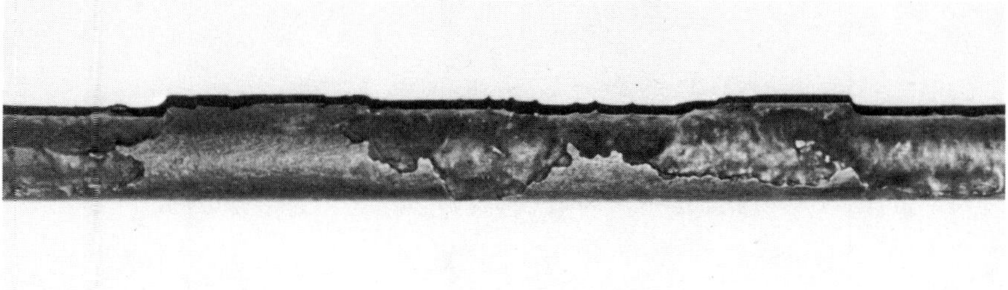

Figure 4–32 CO_2 corrosion of a sucker-rod body showing localized attack and plateaus common to this form of corrosion.

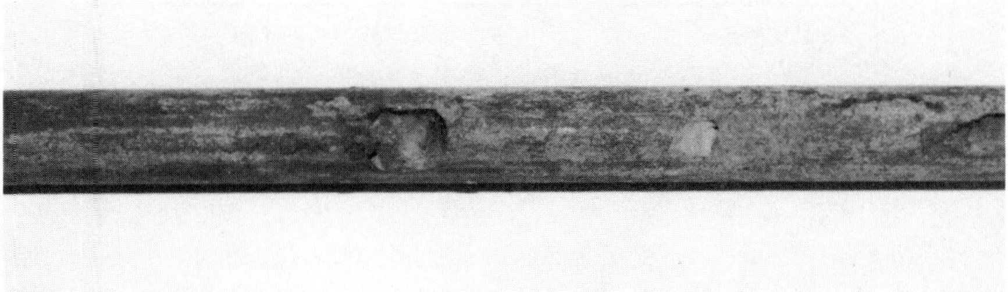

Figure 4–33 Pitting attack of sucker-rod body from H_2S. Note random large pits with little other corrosion of the body.

Figure 4–34 Row of pits along one side of a rod leading to corrosion-fatigue failure.

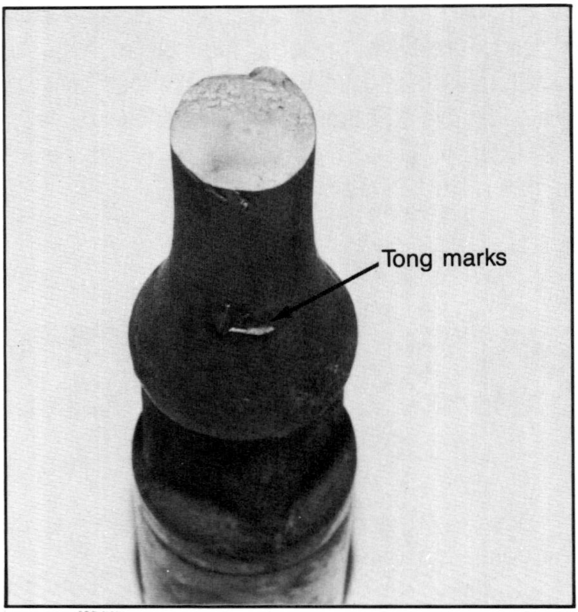

Figure 4–35 Tong marks on shoulder of rod just below coupling that resulted in a fatigue failure.

Mechanical fatigue from improper rod handling is another frequent cause of failure. In this case, nicks and notches from tools induce sufficient stress concentrations to enhance fatigue cracking as discussed in the chapter on drilling. Figures 4–35 and 4–36 show failures related to mechanical damage.

Depending on the environment a sucker rod is subjected to, heat treatment and chemical composition of the rod can have significantly different effects on the fatigue life. In environments in which hydrogen sulfide is present, corrosion fatigue may be accelerated by the effects of sulfide-stress cracking or hydrogen-

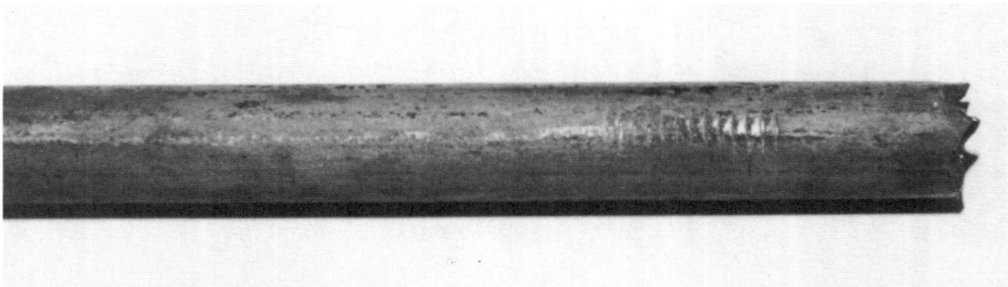

Figure 4-36 Tool marks on body caused notches that ultimately led to a fatigue failure.

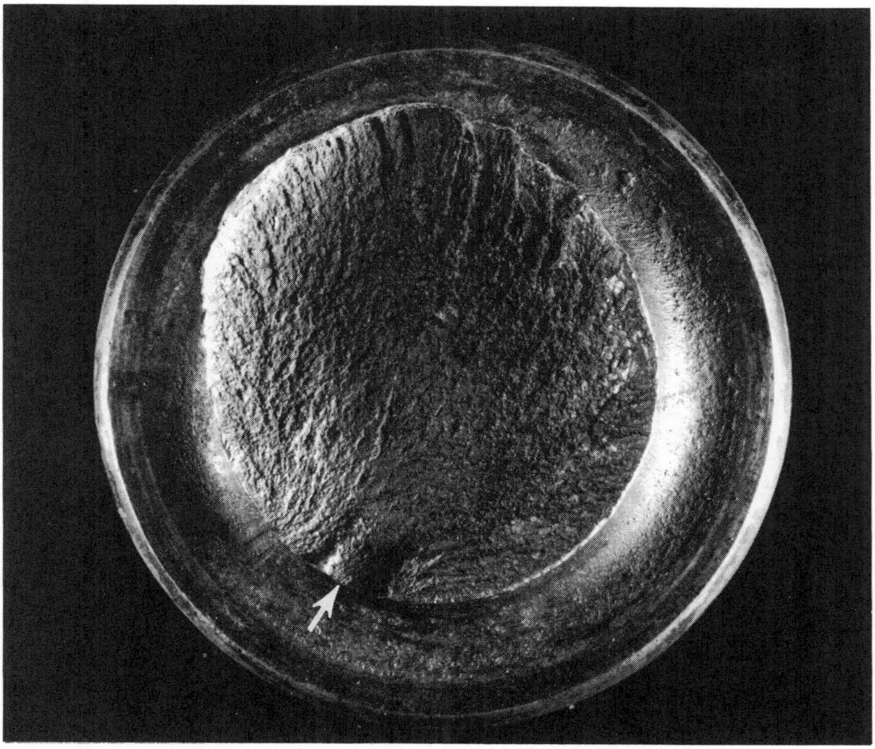

Figure 4–37 Fracture surface of induction-hardened rod from SSC. Arrow indicates fracture origin.

enhanced fatigue-crack growth. Thus, rods with hardnesses above HRC 22 may not perform as well as lower-strength rods. In fact at high strength, rapid fracture by SSC can occur with essentially no fatigue. Figure 4–37 shows an induction-hardened rod that has failed by SSC. Table 4–7 illustrates that higher-strength rods in hydrogen sulfide can give poor performance. The addition of alloying elements such as chromium and molybdenum (e.g., 1036 versus 4142) does not increase service life as might be expected but can act to decrease it. In fact, much work on corrosion fatigue of carbon steels versus low-alloy steels shows that minor amounts of alloying do not increase fatigue resistance, nor do they enhance corrosion resistance. This is because there just is not enough alloy present to provide resistance to pitting, which is one of the initiating factors in corrosion fatigue.

In sweet wells in which only chlorides and carbon dioxide are present, some evidence indicates low-alloy steels may have an advantage over carbon steels (e.g., 1036 steel) from a corrosion-resistance standpoint. Figure 4–38 shows the reduced corrosion of AISI 4340 and 3235 steel compared to 1036 steel in a

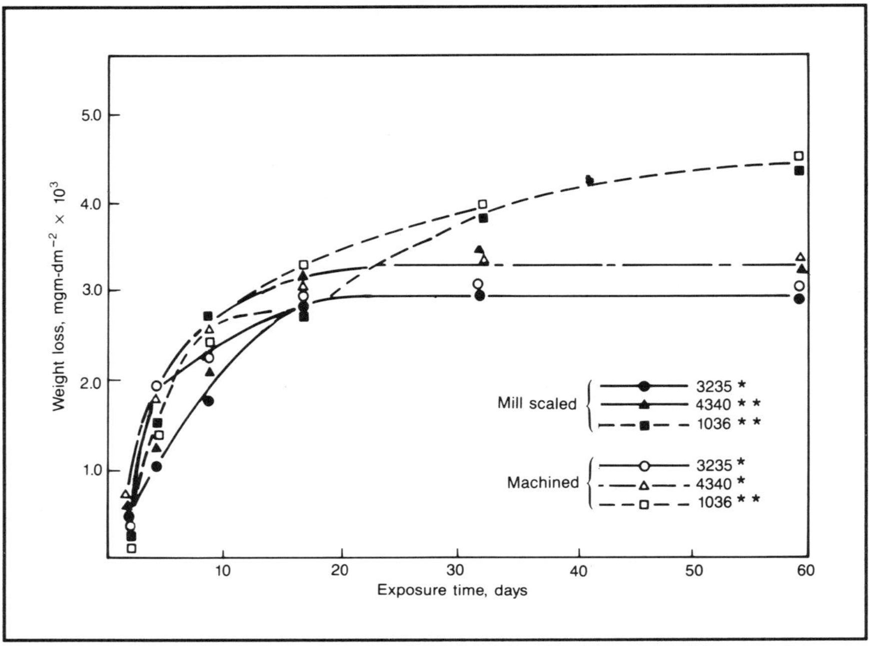

Figure 4–38 Weight-loss data for AISI 3235, 4340, and 1036 Steels. *Hot-rolled normalized and tempered. **Hot-rolled and normalized (After Snape and Van Rooyen, © ASME).

sweet environment. In practice, however, little difference is usually found in the corrosion-fatigue resistance of carbon and low-alloy steels. There are few high-alloy rods available (see Table 4–6), and little information exists on their behavior in corrosive wells. More often fiberglass rods are used for corrosion resistance than are corrosion resistant alloy rods.

One method to promote corrosion resistance and increase corrosion-fatigue life has been plasma-arc spraying of a 316 stainless steel coating on Grade C or D rods. In this process, particles of the stainless steel are carried by an arc-plasma stream to the surface of the rod, where they immediately solidify. This is followed by the application of a fusion bond epoxy top coat. The total coating thickness is typically on the order of 16–17 mils. Laboratory testing (see Figure 4–39) shows the beneficial effect of this process when the coated rods are subjected to an environment containing hydrogen sulfide, carbon dioxide, and brine. The coated rods can be subjected to a much higher peak stress than uncoated rods, and a true threshold behavior returns to the coated rod. However, above 55 ksi peak applied stress, the brittle 316 coating cracks and fatigue resistance is reduced compared to the uncoated rod. Using a fatigue-life criteria

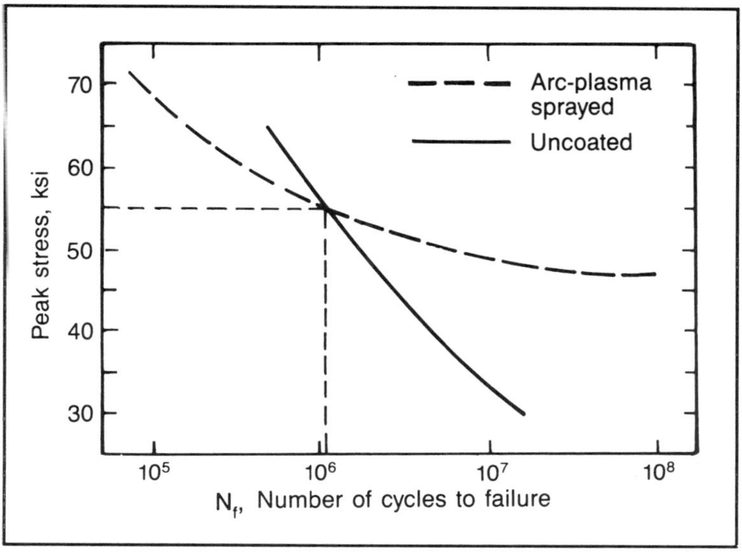

Figure 4–39 Enhancement of fatigue life for arc-plasma-sprayed 316 stainless-steel rods over uncoated Grade C rods (After Chen and Mehdizadeh).

10^7 cycle, it is evident that a 40% increase in peak operating stress is available by using the coated rod. However, the bonding strength of the coating is the limiting factor in performance of plasma-sprayed rods, so care must be taken when running higher-strength rods.

Sucker-rod couplings can be another source of rod string failures. Improper makeup of the couplings to the rods can result in coupling fatigue failures. Looseness of the joint allows fatigue crack initiation at the root of the coupling thread next to the last engaged thread of the pin. A typical fracture surface in a coupling is shown in Figure 4–40. Another frequent cause of coupling failure is externally initiated. For example, hammering the coupling, pipe-wrench jaw marks, corrosion, or wear can produce notches that initiate fatigue cracks from the outside surface of the coupling, as shown in Figure 4–41.

Less often, failure of the pin can occur if the joint is overtorqued. This will usually be quite apparent when the joints are being made up, and failure will occur in the shoulder of the rod.

Wear and corrosion resistance can be dealt with metallurgically by using spray metal couplings. *API Specification 11B* designates two grades of couplings; T (bare metal) and SM (spray metal). The hardness of the former must be in the range HRC 16–23 and the spray metal on the latter greater than HV 595.

The use of bare metal and spray metal coupling and rods is at the center of a

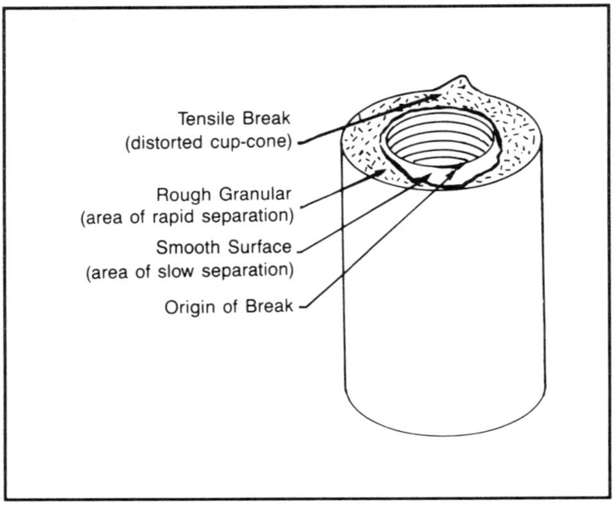

Figure 4–40 Sucker-rod coupling-fatigue failure initiated at ID (After EnDean).

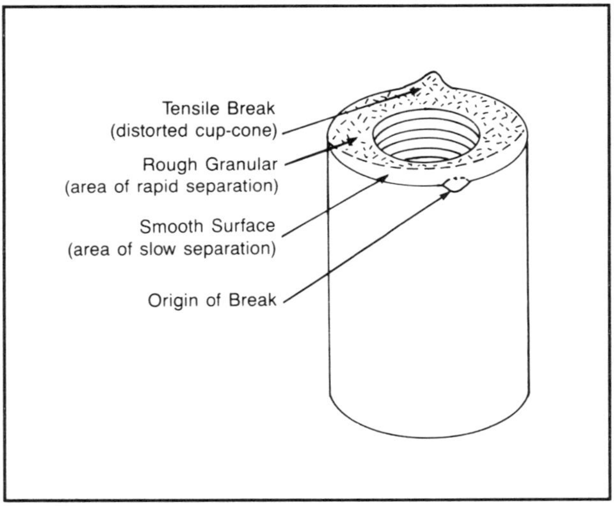

Figure 4–41 Sucker-rod coupling-fatigue failure initiated at OD (After EnDean).

predicament in rod-pumped wells. In many rod-pumped wells for numerous reasons the rods slap against the tubing and/or the couplings rub against the tubing. Either the tubing, rods, or couplings then preferentially fail by rapid metal loss. This is frequently and incorrectly blamed on wear and a hard surfacing or spray metal is applied to the couplings or rods for wear resistance.

Of course this just transfers the problem to the tubing. In actual fact, these problems are not primarily wear or erosion but rather corrosion accelerated by wear. As the rods slap the tubing or the couplings rub against the tubing they locally remove the semi-protective corrosion product layer and/or the corrosion inhibitor leaving bare metal behind that quickly corrodes. After each stroke, more bare metal is available for corrosion in preference to those areas that are still covered with corrosion product and/or inhibitor. Accelerated corrosion then ensues until failure of the rods, couplings, or tubing occurs. Although centralizers and rod guides would reduce tubing wear from couplings and rods, there is a reluctance by workover crews to run this type of equipment. Care should be taken not to create one problem while solving another.

Subsurface Pumps

Although a section on subsurface pumps should include electric submersible pumps, little data are available on these pumps and historically they have not had the number of problems or been exposed to the more severe environments to which sucker-rod pumps have been exposed. Sucker-rod pump requirements are detailed in *API Standard 11AX*, but little information is provided about the metallurgy. Most pumps fail by corrosion and/or abrasion. There are three basic types of subsurface pumps: rod pumps, tubing pumps, and casing pumps. Just as the designators imply, the first operates by inserting the pump into the tubing string and landing it on a holddown. For the second, the pump barrel and standing valve are installed as part of the tubing string, and the third is like the first except it is installed on the sucker-rod string and run in the casing without using tubing. A typical rod pump is shown in Figure 4–42 with the various parts designated.

Table 4–8 gives the chemistries of the most common alloys used for barrels and plungers in sucker-rod pumps. Although this is by no means complete, it does represent a majority of those alloys used in the manufacture of subsurface pumps. Tables 4–9 to 4–13 give recommendations on the applications of barrel tube, plunger, ball and seat, and cage metallurgy for corrosive and erosive service.

Since carbon-steel components corrode rapidly in even slightly corrosive environments containing hydrogen sulfide and/or carbon dioxide, it is common to chrome plate many surfaces, including the pump barrel. However, chromium plate is porous, quite hard, and easily cracked, which leads to a severe galvanic effect at the breaks. The resulting galvanic corrosion can be rapid and catastrophic. Figure 4–43 shows an example of this problem on a pump plunger.

Table 4–8 Classification of Sucker-Rod Steels According to API Specification in Relation to Their Corrosion-Fatigue Performance in $H_2S + CO_2$ + Brine

Material	Heat Treatment	UTS, ksi	Hardness Rockwell C	Corrosion Fatigue Life at 36,000/12,000 psi Stress No. of Cycles x 10^6	Corrosion Fatigue Life at 36,000/12,000 psi Stress Service Life Months[1]	Min. UTS, ksi API Spec	API Class
1036	N	102	8	8.63	13	90	C
4621	N&T (1,000 °F)	89	10	5.19	7.5	82	K
1036	Q&T (1,000 °F)	124	21	9.10	13.2	115	D
1036	Q&T (900 °F)	137	29	4.72	6.8	115	D
4142	N&T (1,000 °F)	127	27	5.96	8.5	115	D
4340	N&T (1,000 °F)	134	28	3.73	5.4	115	D

[1]Equivalent service life of a polished sucker rod at a rate of 16 cpm.
(After Mehdizadeh)

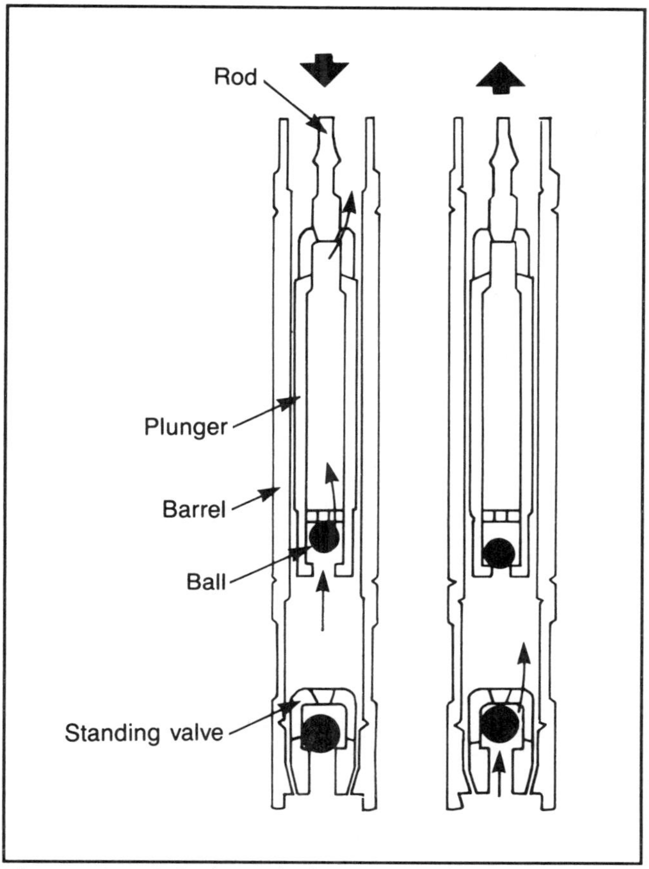

Figure 4–42 Sucker-rod tubing pump showing the different components (After Bruton).

Electroless nickel plate can have exactly the same type of problem. This strong galvanic couple can be overcome by plating over an alloy such as brass, monel, or stainless steel that is closer in the galvanic series. Pump plungers are often sprayed with a material like Colmonoy No. 6 (chromium boride and chromium carbide in a nickel matrix) for wear resistance. More recently titanium boride and silicon carbide coatings have been effectively used.

Wear is one of the most common causes of failure of subsurface pump components, so many of the parts are surface treated with spray metal or by nitriding, carburizing, and plating. Because of this mixed metallurgy and the desire to use as economical a base metal as possible, galvanic corrosion is the primary corrosion problem in rod pumps, although SSC can also be a problem with heat

Table 4–9 Chemical Composition of Most Common Alloys Used for Subsurface Pump Barrels and Plungers

Alloy	C	Mn	Si	Element (wt%, max.) S	P
1015	0.18	0.60		0.05	0.04
1020	0.23	0.60		0.05	0.04
1025	0.28	0.60		0.05	0.04
1035	0.38	0.90		0.05	0.04
1040	0.44	0.90		0.05	0.04
4130	0.33	0.60	0.35	0.04	0.04
4615	0.18	0.65	0.35	0.04	0.04
4640	0.43	0.80	0.35	0.04	0.04
8645	0.48	1.00	0.35	0.04	0.04
Nitralloy N	0.27	0.70	0.40		
Nitralloy135	0.40	0.70	0.40		
4–6 Chrome	0.10	0.60	0.50	0.04	
303 SS	0.15	2.00	1.00	0.15 min.	0.20
304 SS	0.08	2.00	1.00	0.03	0.045
70–30 Brass					
85–15 Brass					
443 Admiralty brass					
Monel R405	0.30	2.00	0.50		
Monel R400	0.30	2.00	0.50		
Gray cast iron	3.30	0.70	2.40	0.12	
White cast iron	3.20	0.60	0.50	0.12	0.50
Eutectic cast iron	3.70	0.60	2.00	0.05	

(After Bruton)

treatments that produce high hardness. Specific information on selection of alloys for sucker-rod pumps is available in *NACE Standard MR-01-76*, "Metallic Materials for Sucker Rod Pumps for Oilfield Corrosive Environments."

Christmas Tree and Wellhead Equipment

The Christmas tree represents the primary surface control of a well. Figure 4–44 shows a typical tree of which there are numerous variations, depending on temperature, pressure, and produced fluids. The two means of flow control are

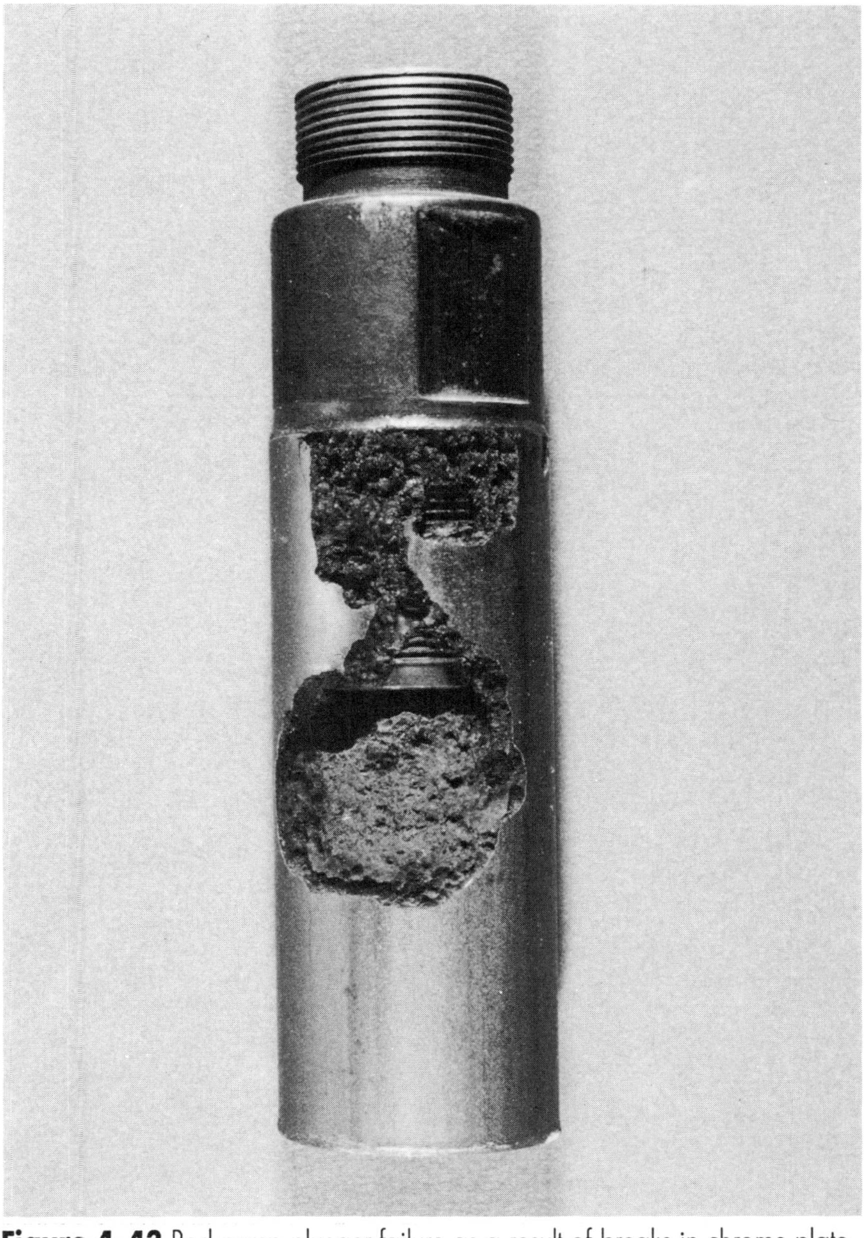

Figure 4–43 Rod pump plunger failure as a result of breaks in chrome plate leading to galvanic corrosion of low-alloy steel plunger body.

Table 4–10 Barrel Tube Recommendations

	1020,1025,1035	C Steel, Cr Plated	1020 Ni Plated	1015, 1020 Carburized	1020 Carbonitrided	1040 Flame Hardened	8645 Cr Plated	4130, 4640 Flame Hardened	Nitralloys, Nitrided	4–6 Chrome
Rockwell Hardness C	**15**	**70**	**68**	**65**	**70**	**56**	**70**	**56**	**70**	**18**
1 None	A	A	A	A	A	A	A	A	A	A
2 None + abrasion	X	A	A	A	A	A	X	A	A	X
3 Severe H_2S	X	X	B	X	C	X	C	C	C	C
4 Severe H_2S + abrasion	X	X	B	X	C	X	X	X	C	X
5 Mild H_2S	C	C	B	X	B	X	B	C	C	C
6 Mild H_2S + abrasion	X	X	B	X	C	X	B	X	C	C
7 Severe CO_2	C	C	B	X	C	X	C	X	X	B
8 Severe CO_2 + abrasion	X	C	B	X	C	X	X	X	X	X
9 Mild CO_2	B	B	B	X	C	X	C	X	X	B
10 Mild CO_2 + abrasion	X	B	B	X	C	X	C	X	X	X
11 Severe H_2S + CO_2	X	X	B	X	C	X	C	X	X	C
12 Severe H_2S + CO_2 + abrasion	X	X	B	X	C	X	X	X	X	X
13 Mild H_2S + CO_2	X	X	B	X	C	X	C	X	X	C
14 Mild H_2S + CO_2 + abrasion	X	X	B	X	C	X	X	X	X	X
15 Severe brine	B	C	A	B	B	C	B	C	B	B
16 Severe brine + abrasion	X	C	A	B	B	C	B	C	B	X
17 Mild brine	B	B	A	A	A	B	B	B	A	A
18 Mild brine + abrasion	X	B	A	A	A	B	B	B	A	X
19 Dissolved oxygen	B	B	A	C	C	X	B	X	B	B

A = Material suitable
B = Corrosion or erosion expected
C = Corrosion or erosion severe; not recommended
X = unsuitable
(After Bruton)

valves and chokes. Most often, Christmas-tree valves (e.g., wing, master, and swab) are gate-type valves. Since chokes are of similar design, the following discussion will refer to general aspects of both because the metallurgical requirements for fracture and corrosion resistance will be essentially identical. Christmas-tree components are covered by *API Specification 6A*. Most valve bodies to be used in noncorrosive environments at temperatures from -20 to

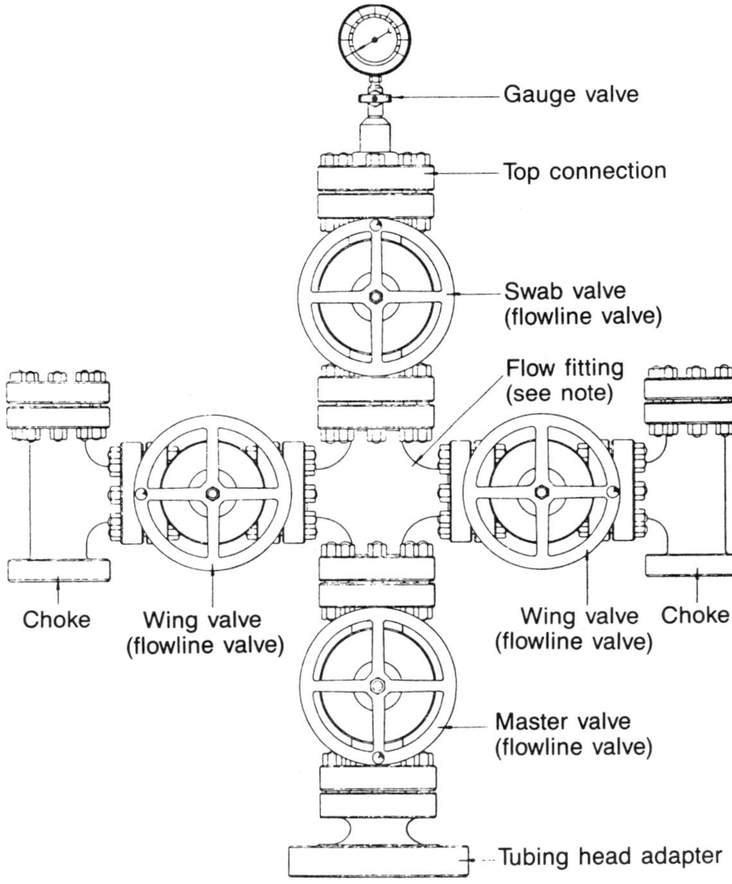

Figure 4–44 Typical Christmas tree (After *API Specification 6A*).

250°F are constructed of ASTM A 487 Grade 4 steel, AISI 4130, 4140, or 8630 steel, either forged or cast. The bonnets are made from wrought material of the same composition, usually 4140 steel, as are the stem and gates. Seats typically are also made of 4140 steel, although there are some variations from manufacturer to manufacturer. Gates are usually molydisulfide coated or plated with nickel or chromium.

For application in corrosive environments, a variety of metallurgy is available from manufacturers depending on the particular conditions of the well.

Table 4–11 Plunger Recommendations

	1015,1035,1040	Hardened	Chrome Plated	Chrome Plated — Molykote	Nickel Plated	Spray Metal	Spray Metal, Threads Protected	4615, 4-6 Cr	Hardened
Rockwell Hardness C	**15**	**56**	**70**	**70**	**68**	**62**	**62**	**20**	**56**
1 None	A	A	A	A	A	A	A	A	A
2 None + abrasion	X	B	A	A	B	A	A	X	B
3 Severe H_2S	X	X	X	X	B	C	A	X	X
4 Severe H_2S + abrasion	X	X	X	X	B	C	A	X	X
5 Mild H_2S	X	X	X	X	A	A	A	C	X
6 Mild H_2S + abrasion	X	X	B	B	A	A	A	X	X
7 Severe CO_2	X	X	X	X	B	X	A	X	X
8 Severe CO_2 + abrasion	X	X	X	X	B	C	A	X	X
9 Mild CO_2	X	X	B	B	A	A	A	X	X
10 Mild CO_2 + abrasion	X	X	B	B	A	A	A	X	X
11 Severe H_2S + CO_2	X	X	X	X	B	X	A	X	X
12 Severe H_2S + CO_2 + abrasion	X	X	X	X	B	X	A	X	X
13 Mild H_2S + CO_2	X	X	X	X	B	X	A	X	X
14 Mild H_2S + CO_2 + abrasion	X	X	X	X	A	B	A	X	X
15 Severe brine	X	X	X	C	A	B	A	X	X
16 Severe brine + abrasion	X	X	C	C	A	B	A	X	X
17 Mild brine	C	C	C	B	A	A	A	C	C
18 Mild brine + abrasion	X	C	C	B	A	A	A	X	C
19 Oxygen	C	X	C	C	A	B	A	C	X

A = Material suitable
B = Corrosion or erosion expected
C = Corrosion or erosion severe; not recommended
X = unsuitable
(After Bruton)

When SSC is a potential problem and NACE MR-01-75 requirements are mandated, the body and bonnet are often made of 8630, 4140 or 2-1/4 Cr-1 Mo quenched and tempered to a maximum hardness of Rockwell C 22. The gate and seats are made of Type 410 or 17-4PH stainless steel. Sometimes seats are overlayed with Stellite No. 6 for wear and corrosion resistance. Stems are also made of 410 stainless steel, Monel K500, or 17-4 PH precipitation hardening) stainless steel. In very sour service all of these stem materials have failed from

Table 4–12 Ball and Seat Recommendations

Type	Maximum Chemistry (Wt%)						
Steel	**Cr**	**Ni**	**C**	**Si**	**Mo**	**S**	**Se**
Ball—tool steel	11.50		1.55		0.80		
Seat—tool steel	11.50		1.55		0.80		
Stainless Steel	**Cr**	**Ni**	**C**	**Si**	**Mo**	**S**	**Se**
Ball—440C	18.00		1.20	1.00	0.75		
Seat—414C	13.50	2.00	0.15	1.00		0.03	
Ball—440C	18.00		1.20	1.00	0.75		
Seat—440A	18.00		0.70	1.00	0.75	0.03	
Ball—329	27.50	4.50	0.08		1.50		0.85
Seat—440A	18.00		0.70	1.00	0.75	0.03	
Nonferrous	**Ni**	**Cr**	**Cu**	**C**	**Mn**	**Al**	**Ti**
Ball—K-500 monel	70.00		Bal	0.25	1.50	3.15	0.85
Seat—K-500 monel	70.00		Bal	0.25	1.50	3.15	0.85
Ball—bronze			80.00		0.50	14.00	
Seat—bronze			80.00		0.50	14.00	
Cast cobalt alloy	**W**	**Co**	**Cr**	**C**	**Mn**	**Mo**	**Fe**
Ball—Rexalloy 35	18.00	Bal	33.00	2.25			
Seat—Rexalloy 44	8.75	Bal	32.75	1.85	1.00	0.50	1.00
Ball—Rexalloy 35	18.00	Bal	33.00	2.25			
Seat—Stoody	13.00	Bal	26.00	2.50	0.50		2.00
Sintered Carbide	**W**	**Co**	**Ni**	**Cr**	**C**	**Ti**	**—**
Ball—chrome carbide	2.00	10.00	15.00	83.00			
Seat—tungsten carbide	65.00	8.50			7.60	11.00	
Ball—titanium carbide	4.50		25.00			52.00	
Seat—tungsten carbide	65.00	8.50			7.60	11.00	
Ball—tungsten carbide	65.00	8.50			7.60	11.00	
Seat—tungsten carbide	65.00	8.50			7.60	11.00	
Combinations	**W**	**Co**	**Al**	**Cr**	**C**	**Ti**	**Cu**
Ball—bronze			14.00				80.00
Seat—440A-SS				18.00	0.70		
Seat—440C-SS				18.00	1.20		
Seat—tungsten carbide	65.00	8.50			7.60	11.00	

A =Material suitable
B = Corrosion or erosion expected
C = Corrosion or erosion severe; not recommended

SSC. Therefore, for critical sour service stems are made from Inconel 718 or Inconel 625.

In environments in which only carbon dioxide is present but in concentrations expected to be quite corrosive (5% or 30-psig carbon dioxide partial pressure),

Table 4–12 Cage Material Recommendations

Cage and Fitting Mat'l	Material Maximum Chemistry (%)						
	C	Mn	Si	S	Cr	Ni	Mo
Carbon Steel							
1 B-1112	0.13	1.00		0.23			
2 B-1113	0.13	1.00		0.33			
3 B-1137	0.39	1.65		0.13			
4 B-1141	0.45	1.65		0.13			
Low-alloy Steel							
1 4140	0.43	1.00	0.35	0.04	1.10		0.25
2 450	0.53	1.00	0.35	0.04	1.10		0.25
3 4820	0.23	0.70	0.35	0.04		3.75	0.30
4 8620	0.23	0.90	0.35	0.04	0.60	0.70	0.25
Stainless Steel							
1 303	0.15	2.00	1.00	0.40	19.00	10.00	0.60
2 203	0.07	6.50		0.35	17.00		6.00
3 416	0.15	1.25	1.00	0.15	14.00		0.60
Nonferrous	**Cu**	**Zn**	**Ni**	**C**	**Mn**	**Al**	**Others**
1 Monel R405	Bal		70.00	0.30	2.00		Si 0.50
2 Monet K500	Bal		70.00	0.25	1.50	3.15	Ti 0.85 Si 0.50
3 70-30 Brass	0.71	0.31					
4 85-15 Brass	0.86	0.15					
5 Admiralty brass	0.71	0.28					Sn 1.00 As 0.04

Notes: [1]Recommendations for abrasion services based on use of hard-faced or insert-type cage.
[2]Boron—0.0005 min.
A = Material suitable
B = Corrosion or erosion expected
C = Corrosion or erosion severe; not recommended
X = Unsuitable
(After Bruton)

Type 410 stainless steel, ASTM A217 CA-15 (cast stainless equivalent to wrought 410). F6NM (forged equivalent to 410) or CA6NM (ASTM A487) are used for body and bonnet. Type 410 stainless-steel gates and seats are also installed. The gates are often coated with molydisulfide and/or are hard surfaced or nitrided since carbon dioxide is notorious for increasing corrosion in areas of high wear or velocity. Stems are made of either 410 stainless steel or 17-4 PH. In high CO_2 concentration, solid 316 stainless steel trees have been successful.

If both hydrogen sulfide and carbon dioxide are present but pressure is low, the body and bonnet are usually made of 410 stainless steel, (CA-15 or CA6NM), with 410 stainless steel or Monel K500 gates and seats and either Monel K500 or 17-4 PH stems. All parts have controlled hardness conforming to NACE MR-01-75.

For low-temperature applications such as arctic service, notch toughness is the most important criteria. To meet API 6A impact requirements, alloys such as 8630, 4130, 2-1/4 Cr-1 Mo, 410 stainless steel, or 17-4 PH stainless quenched and tempered must be used for all components pressure containment. In cases in which hydrogen sulfide and carbon dioxide or a combination of the two are present, 410 stainless steel and/or 17-4 PH are used. For bodies and bonnets, CA6NM also has found application for low-temperature service because of its good fracture toughness and ability to be heat-treated.

Even with the wide selection of materials available for aggressive environments, these low-alloy steels and stainless steels are not adequate for many highly corrosive environments. As shown in Figure 4–15, 13% Cr stainless steel offers limited resistance to elevated temperature carbon dioxide corrosion. For higher concentrations of carbon dioxide and associated brines at elevated temperature, more protection is needed. In some fields that produce carbon dioxide, type 316 stainless steel bodies and bonnets have been successful because of high velocities through the tree, and temperatures have been low enough to avoid stress-corrosion cracking. However, the achievable strength of austenitic stainless steel components is low, so when high-pressure wells or very corrosive wells are completed, either a heat-treatable CRA must be used for components or corrosion resistant overlays must be applied. The former is often cost prohibitive in most cases although solid duplex stainless steel and Incoloy 925 trees have been made and installed. A more cost effective solution to the use of solid CRA components is to clad the wetted areas with CRA but use a heat-treatable low-alloy steel shell.

Weld overlaying of 4130 or 2-1/4 Cr-1 Mo steel base metal is performed using gas metal-arc welding (GMAW) more commonly known as metal inert-gas (MIG) welding with Inconel 625 wire. All surfaces that will be wetted by the corrosive stream are overlayed and then ground smooth. This type of procedure has several built-in problems that must be examined for each application. For instance, weld dilution will form an alloy at the steel/overlay interface with properties entirely different from either the base metal or welding wire. The heat-affected zone can also become quite hard and susceptible to hydrogen-stress cracking or reduced fracture toughness. If overlaying is incomplete, small anodes of low-alloy steel will be coupled to a large cathode of alloy metal that presents a serious galvanic-corrosion problem. In contrast to these potential drawbacks, weld overlay trees have successfully functioned in very severe environments for over 10 years with no reported failures.

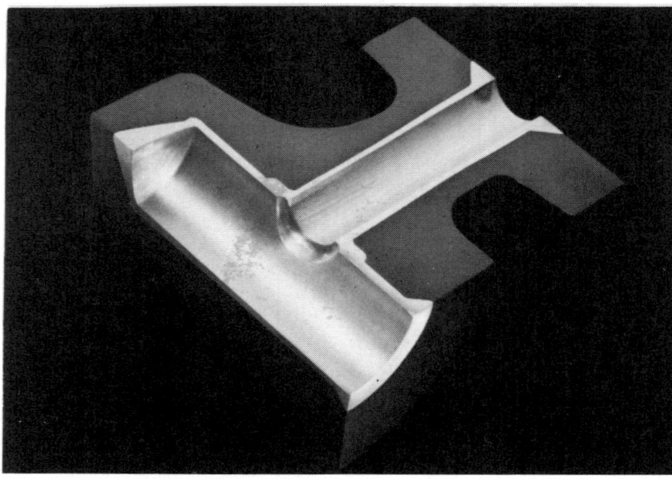

Figure 4–45 Section through a CRA clad valve body.

An alternative to weld overlay is to hot isostatically pressed (HIP) nickel-base alloys into low-alloy bodies. The application of the HIP process to valve bodies allows development of a good metallurgical bond with the base metal and with the assurance of uniform microstructure, properties, and thickness of the CRA. Almost any alloy can be HIPed, and work has been done on a variety of common engineering alloys such as Inconel 625, Hastelloy C-276, and MP35N. Inconel 625 is the most common material for HIP clad and these components have been as successful as the weld overlay components.

The advantage of the HIP-clad valves is the resultant wrought structure of the alloy material, which is more homogeneous and thus less susceptible to localized corrosion as is the cast structure of a weld overlay. Furthermore, interfacial dilution extends only about 0.05 mm into the clad, whereas on overlays it extends on the order of 3.2 mm. However, problems have been encountered in the past with incomplete removal of the steel cans necessary to the HIP process. Remnants of steel on the surface of the CRA can compromise the performance of the clad. Figure 4–45 shows a section of a clad valve body.

The resistance to localized attack in a CO_2-Cl environment as a function of molybdenum content of an alloy is shown in Figure 4–46. These alloys were examined for potential cladding in valves. As is evident, increasing resistance to pitting, which is described by the breakdown potential, occurs with increasing molybdenum content. The very high resistance to pitting of Inconel 625 explains why it is preferred for cladding.

Failures of valve components in hydrogen sulfide systems occur in stem mate-

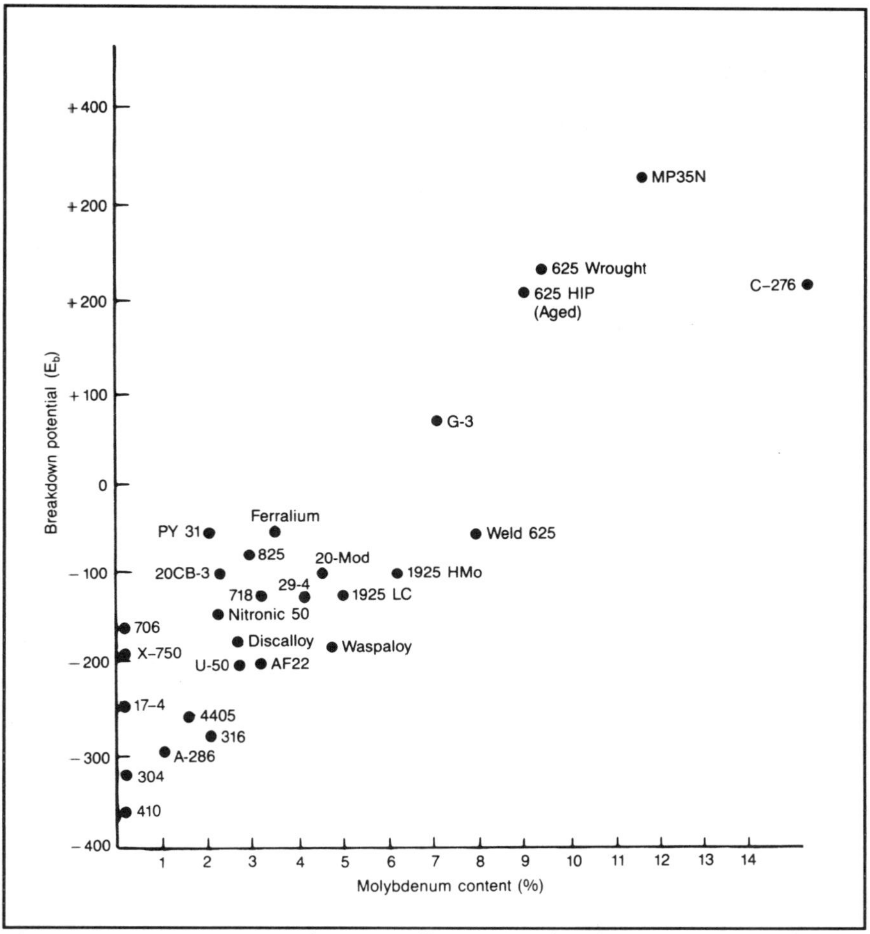

Figure 4–46 Resistance of various alloys to localized corrosion (E_b) as a function of molybdenum content in 11 Wt% NaCl brine with 500 psi of CO_2 at 250°F (121°C) (After Bednarowicz).

rial of hardness that meet those recommended by the NACE MR-01-75 specification. Inconel X-750 was a common alloy used for stem construction until frequent field failures caused by SSC required its replacement with Inconel 718, which has performed much better. Valve stems made of Monel K500 that were heat treated in accordance with MR-01-75 have also failed by SSC at hardnesses below the HRC 35 maximum imposed by MR-01-75. In addition, valve bodies made of 410 stainless steel with hardness greater than HRC 22 and bonnets made from Ledloy A at a hardness of BHN 174 have failed in hydrogen sulfide environments. Neither of these, however, met the MR-01-75 requirements.

Few valve body failures have occurred because of environmental cracking, probably because body and bonnet sections are quite thick for pressure containment; therefore, applied stresses are lower, which reduces the potential for cracking. Bodies more often fail from corrosion in stagnant areas that allow hydrogen sulfide, carbon dioxide, and/or oxygen to accumulate.

Chromium or nickel plating (electrolytic or electroless) is not always beneficial. Gates are often plated for corrosive production service or waterflood service. If the plate becomes damaged or microcracks form, as is often the case in high flow areas, severe pitting attack follows at the base of the crack—again caused by a galvanic effect when the base metal is a low-alloy steel. Systems containing dissolved oxygen or carbon dioxide are particularly susceptible to galvanic attack.

In those cases where corrosion resistance must be combined with good erosion resistance, hard facing using plasma transferred arc (PTA)-applied Tribaloy, or Stellite coatings such as T-800 and No. 4, respectively, are suitable. Also, detonation ("D") gun-applied LW45 has been successfully used for severe service. Selection of a corrosion-erosion resistant hard facing for gates and slats is best accomplished by testing the product in the actual or simulated environment.

One other area in which valves are commonly used is in waterfloods. Here, water is injected into the ground for secondary recovery of oil, and it is common for injection pressures to reach 4,000 psi. Since the water is gathered from a source well or surface source such as tanks, oxygen is a prime culprit in corrosion. If produced water is commingled with other water for injection, hydrogen sulfide and carbon dioxide may also be present. Chlorides and suspended solids, which may impart a high degree of abrasion and erosion, especially to valve components, are usually added to these corrodents.

Bodies and bonnets for waterflood valves are constructed of quenched and tempered 4130 or 4140 steel, while gates are made variously of chromium or nickel-plated low-alloy steel such as 4130, 316 stainless steel, or 17-4 PH stainless steel. Stems are usually made of 17-4 PH stainless steel in the H1150 condition, while seats are made of 316 or 17-4 PH stainless steel.

Most problems in waterflood service are corrosion of the body and gates, erosion of the seats, and erosion-corrosion of the body. Plated gates have short lives if the water is oxygenated—again from galvanic corrosion between the noble metal plating and the low-alloy steel base metal. The same occurs in the body and bonnet when they are plastic coated, usually with a baked phenolic. Local breaks in the coating result in failure because of pitting of the base metal.

The NACE has compiled tables for easy access to information on the metallurgy of valves and components to be used in waterflood service. The pertinent tables are reproduced here as Tables 4–14 to 4–16. The particular limitations and uses of these tables are described in detail in *NACE Recommended Practice*

Material Code for Tabels 4–14, 4–15, and 4–16

Fe	Carbon or low-alloy steel
Fe_c	Carbon or low-alloy steel, coated or lined internally
410	AISI 410 stainless steel (wrought) or A296 grade CA15 (cast)
316	AISI 316 Stainless steel (wrought) annealed or A296 grade CF-8M(cast) annealed
17-4	17-4 PH stainless steel (wrought) orACI grade CB-7Cu (cast)
$NiAl_{ca}$	Nickel aluminum bronze, B148, alloy 955 (9D) (cast)
$NiAl_w$	Nickel aluminum bronze, A150, alloy 630 (9D) (wrought)
400M	Monel 400 (wrought) or Monel 410 (cast)
K500	Monel K500 (wrought) or Monel 505 (cast)
In_{all}	Inconel alloys (all listed grades)
In750	Inconel X–750
WC	Tungsten carbide
HardNi	Ni–Cr–B–Si alloys (such as Colmonoy)
Hard Co	Cobalt-base alloys (such as Stellite)
Ni	Nickel plating
NM	Nonmetallic

RP-04-75 and should be consulted prior to any selection. The tabulated conditions are for aerated water containing more than 10 ppb oxygen: nonaerated, less than 10 ppb; with H_2S, more than 1 ppm, and without H_2S, less than 1 ppm.

One other component of Christmas trees that frequently require special metallurgy are chokes. Chokes are frequently subjected to erosion from liquids or solids in the flow stream, abrasion and/or cavitation. Abrasion is associated with surfaces sliding against each other and is accelerated by particles between these surfaces. Cavitation is wear damage caused by the formation and collapse of vapor bubbles in a fluid. These bubbles can create pressures on the order of 200,000 psi when they collapse, severely damaging the metal surface. Thus chokes are subjected to extremely severe wear and the choice of materials to resist these forms of wear are not always obvious. For example, increasing hardness does not always translate to increasing erosion resistance since there is often a corrosion component to the attack; therefore, the most resistant material may be softer than other alloys but more corrosion resistant. It has been found that for excellent erosion and abrasion resistance, tungsten carbide with very low nickel or cobalt binder content and ceramics are the best. For cavitation resistance in liquid service, stellites and nickel chromium alloys followed by stainless steels, tungsten carbide, and ceramics are best.

Table 4–14 High- and Low-Pressure Discharge Gate Valves

	Environment			
	Aerated		Nonaerated	
	Without H_2S	With H_2S	Without H_2S	With H_2S
Body & bonnet	316, $NiAl_{ca1}$, $NiAl_w$ Fe_c	316, $NiAl_{ca1}$, $NiAl_w$ Fe_c	316, $NiAl_{ca1}$, $NiAl_w$ Fe_c	316, $NiAl_{ca1}$, $NiAl_w$ Fe_c
Stem	In750 K500 400M[1], 316[2] 17–4	In750 K500[4] 316 17–4	In750 K500 17–4	In750 K500[4] 17–4
Gate	(WC or HardCo/316, HardNi)/316 400M[3]/Ni 316/[3]Ni $NiAl_{ca}$	(WC or HardCo/316, HardNi)/316 400M[3]/Ni 316/[3]Ni $NiAl_{ca}$	(WC or HardCo/316, HardNi)/316 400M[3]/Ni 316/[3]Ni $NiAl_{ca}$	(WC or HardCo/316, HardNi)/316 400M[3]/Ni 316/[3]Ni $NiAl_{ca}$
Seats	In_{all}, 400M, 316 HardCo, HardNi	In_{all}, 400M, 316 HardCo, HardNi	In_{all}, 400M, 316 HardCo, HardNi	In_{all}, 400M, 316 HardCo, HardNi
Carrier nut	316, In_{all}, 400M 410 $NiAl_{ca}$, $NiAl_w$	316, In_{all}, 400M $NiAl_{ca}$, $NiAl_w$	316, In_{all}, 400M 316, 410 $NiAl_{ca}$, $NiAl_w$	316, In_{all}, 400M 410 $NiAl_{ca}$, $NiAl_w$

[1]Low pH (>3, <6)
[2]High pH only (>6)
[3]/Means overlay
[4]Must conform to NACE *Standard MR-01-75*
(After NACE)

Table 4–15 Throttling Valves

	Environment			
	Aerated		Nonaerated	
	Without H_2S	With H_2S	Without H_2S	With H_2S
Body & bonnet	316, 316L Fe_c	316, 316L Fe_c	$NiAl_w$[1], 316[2], 316L[2] Fe_c	$NiAl_w$[1], 316[2], 316L[2] Fe_c
Stem	316, K500 410[5]	316, K500[6]	316, K500	316, K500[6]
Gate	HardCo, HardCo/[3]	HardCo, HardCo/[3]	HardCo, HardCo/[3]	HardCo, HardCo/[3]
Disc	HardCo, 316	HardCo, 316	HardCo	HardCo
Carrier nut	316, In_{all}, 400M 410[4] $NiAl_{ca}$	316, In_{all}, 400M 410[4] $NiAl_{ca}$	316, In_{all}, 400M 410[4] $NiAl_{ca}$	316, In_{all}, 400M 410[4] $NiAl_{ca}$

[1]Not in common use
[2]Limited experience
[3]/Means overlay
[4]Widely used material
[5]Fresh water only
[6]Must conform to NACE *Standard MR-01-75*
(After NACE)

Table 4–16 Well Equipment

	Environment			
	Aerated		Nonaerated	
	Without H_2S	With H_2S	Without H_2S	With H_2S
Supply wells				
Casing head	Fe	Fe	Fe	Fe
Tubing head	Fe_c, NM	Fe_c, NM	Fe, Fe_c, NM	Fe_c, NM
Tubing hanger				
Exposed	Fe_c	Fe_c	Fe_c	Fe_c
Nonexposed	Fe	Fe	Fe	Fe
Tubing nipples	Fe_c, NM	Fe_c, NM	Fe_c, NM	Fe_c, NM
Tubing	Fe_c, NM	Fe_c, NM	Fe_c, NM, Fe[1]	Fe_c, NM
Packers	Fe_c	Fe_c	Fe_c, Fe[1]	Fe_c
Injection wells				
Tubing head	$NiAl_{ca}$, 316, Fe_c	$NiAl_{ca}$, 316, Fe_c	$NiAl_{ca}$, 316, Fe_c, Fe[1]	$NiAl_{ca}$, 316, Fe_c
Tubing hanger	Fe_c	Fe_c	Fe_c	Fe_c
Tubing nipples	$NiAl_w$[2], 316[2], Fe_c	$NiAl_w$[2], 316[2], Fe_c	$NiAl_w$[2], 316[2], Fe_c, Fe[1]	$NiAl_w$[2], 316[2], Fe_c
Packers				
Supply well				
(in parts)	Fe_c	Fe_c	Fe_c	Fe_c
Injection wells				
(all parts)	Fe_c	Fe_c	Fe_c	Fe_c

[1]High pH only (>6)
[2]Where alloy tubing nipples are connected to Fe or Fe_c, the nipple shall be coated
(After NACE)

Wirelines

Wirelines are used for servicing downhole equipment, setting downhole valves, and running measurement tools such as calipers and logging tools. Wirelines are covered by *API Specification 9A* and are constructed of plow steel, improved plow steel, or extra-improved plow steel and coated with zinc by electrodeposition or hot galvanizing.

Plow steel has a composition similar to AISI 1067 steel and has a tensile strength exceeding 240,000 psi; improved plow steel, which is produced from AISI 1075 steel, has a strength of 250,000 psi or higher. This high strength is achieved by heat treating and cold drawing in a process termed patenting.

Heat treatment and cold work produce a very fine pearlitic microstructure that is aligned in the lengthwise direction of the cable, imparting a very high yield and tensile strength to the cable as well as a high degree of anisotropy. These cables are also treated with a corrosion-inhibitor compound that is effective in many environments. However, this type of treatment, as well as wellsite chemical inhibition, is not totally effective against corrosion, and environmental cracking can often occur, especially in hydrogen sulfide-containing environments.

Monel K500 wirelines have been successfully run in western Canada, but in areas of high hydrogen sulfide and high temperature, dealloying can be significant.

Since the diameter of most wirelines is between 1.68 and 2.34 mm, a small reduction in cross section caused by corrosion can produce failure. Type 410 stainless steel has been used; but because of the high strength required and the stresses on the wireline, it too fails in hydrogen sulfide-bearing environments. The austenitic stainless steels, when highly cold worked, are susceptible to chloride stress-corrosion cracking as well as sulfide-stress cracking; therefore, they are not good candidates for this type of service. Since almost all oil and gas wells contain water with chlorides, the use of chloride cracking-sensitive materials should be excluded from wireline service. Some wirelines made of austenitic stainless steels have failed by chloride stress-corrosion cracking when temperatures exceeded 180°F. The author has analyzed failures of a chromium-modified Hadfield manganese-steel wireline (11% Cr, 8% Mn, 2% Ni and 0.80% C) that failed by SSC in a sour well in Wyoming.

In fact, because of the very high tensile strengths required for wireline wire, few alloys are suitable for service in the presence of chlorides and hydrogen sulfide. This is evident by the data presented in Table 4–17 from work by Vaughn and Chaung. The tensile strength of all of the alloys tested was approximately 220 ksi. Under the most severe circumstances—a laboratory solution—the only alloy resistant to environmental cracking was MP35N. This is summarized in

Table 4–17 Stress-Corrosion Cracking of Wireline Materials[1]

Materials	Lab (93°C)	Field (93°C)	Lab (149°C)	Field (149°C)	Lab (177°C)	Field (204°C)
AISI 316	F	F	F	F	. . .	F
18-18-2	F	P	F	P	. . .	F
Uranus 50	NF	NF	F	F	F	F
SAF 2205	NF	. . .	F	. . .	F	F
Uranus B6	NF	NF	F	F	F	F
2RK65 (Sabdvik)	NF	NF	F	NF	F	F
Sanicro28	NF	. . .	F	. . .	F	F
Carpenter 20Cb-3	NF	. . .	NF	NF	F	F
Incoloy 825	NF	. . .	NF	. . .	F	F
Hastelloy G	NF	. . .	NF	. . .	. . .	F
Alloy 718	. . .	NF	NF	NF	. . .	F
Pyromet 31	NF	. . .	F**	. . .	F	F
Inconel 625	. . .	. . .	NF	NF	. . .	F
Inconel 617	. . .	. . .	NF	NF	F	F
Hastelloy C276	. . .	NF	NF	NF	F	F
MP35N	. . .	. . .	NF	NF	NF	NF

[1]Environment —Field (NaCl + $MgCl_2$ + $CaCl_2$ with 11% H_2S and 40% CO_2 gas for 30 days)
—Lab (NaCl with 100% H2S gas for 7 days)

NF = No failure
F = Failure
P = Pitted but no failure
F** = three of six specimens failed
(After Vaughn and Chaung)

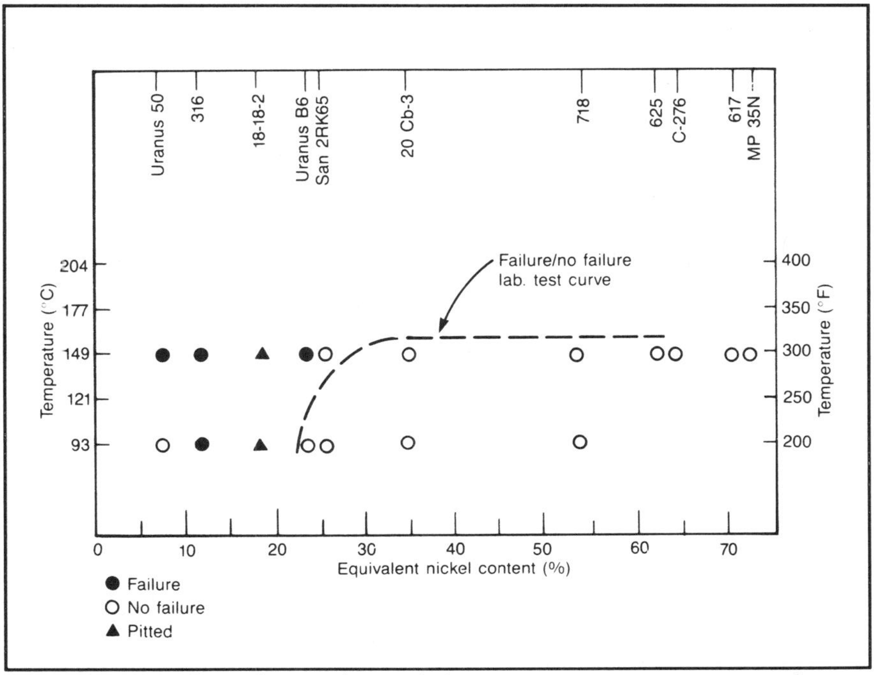

Figure 4–47 Effect of equivalent nickel content on stress-corrosion cracking resistance of wireline materials in field environments (After Vaughn and Chaung.)

Figure 4–47, which relates the cracking sensitivity to nickel content. This behavior explains why MP35N is currently being used for wirelines in extremely severe service and why, to date, no failures of MP35N wirelines have been reported.

The conductor material in these MP35N armored cables is usually nickel-clad copper, ASTM B-355 class 10. Other armor materials currently available are 18-18-2, Uranus 50 (a duplex stainless steel), and Uranus B6, which is fully austenitic. All three alloys are described in Table 4–17 and show less-satisfactory performance than MP35N. At lower temperatures such a 93°C, these should be acceptable alloys and more economical than MP35N.

Miscellaneous Equipment

Although there is a variety of ancillary equipment for the production of oil and gas—far too many to discuss—there are more familiar items that are at least worth mentioning in regard to their metallurgy.

Gas lift mandrels of the side pocket type are made from AISI 4130 steel tube with a welded-in side pocket. The whole assembly is then quenched and tempered to either a J-55 or N-80 grade.

Endless or coiled tubing used for well cleanup, acid stimulation, etc., is commonly 1-in. seam-welded tube with a composition similar to AISI 1015 steel modified with 0.50 Cr. These tubes most often fail because of tensile overload but have also been known to succumb to SSC. Most often in acid work, inhibitors are run with the acid to prevent weight-loss corrosion and pitting attack since the tube wall is normally quite thin.

Gravel-pack screens are used to control sand production that eventually leads to erosion. Figure 4–48 shows the cross section of a screen used for corrosion testing that is a segment of an actual screen. The inner tube is perforated J-55, K-55, or N-80 grade pipe with the outside wire forming the screen being 304 stainless steel, 316L stainless steel, Carpenter 20, or Monel 400. These are spot welded to runners on the base tube. The most common problem with these screens is simply erosion of the fine wires, which leads to large holes in the screen.

One other failure mode is stress-corrosion cracking (SCC)caused by the completion fluids in the wellbore. $CaBr_2$, $CaCl_2$, and $ZnBr_2$ are used to complete wells, and at high temperatures (≥200°F), they can cause SCC. This is especially true for the $ZnBr_2$/304 stainless-steel combination.

References

American Petroleum Institute. *Specifications 5CT, 6A, 9A,* and *11B.*

American Society for Metals. *Metals Handbook.* Volume 1. 1961.

Battle, J.L.; T.V. Miller; and M.E. True. "Resistance of Commercially Available High Strength Tubular Goods to Sulfide Stress Cracking." *Materials Performance,* 14 (1975), p. 11.

Bednarowicz, T.A. "Electrochemical Polarization to Determine the Localized Corrosion Behavior of Various Materials Used for Heavy Section Wellhead Equipment in a 121°C-3.5 MPa CO_2-NaCl Environment." Corrosion/82, Houston, Texas: NACE, 1982.

Brennan, J.R. "Metallurgy of Sucker Rod Pumps." Paper presented at *Western States Corrosion Seminar.* San Dimas, California: NACE, May 1968.

Bruton, B.R. "Selection of Metallic Materials for Subsurface Pumps for Various Corrosive Environments." *Proceedings of the University of Oklahoma Corrosion Control Course.* University of Oklahoma and NACE, September 1971.

Chen, E.Y. and P. Mehdizadeh. "Corrosion Fatigue Life of Arc Plasma Stain-

Figure 4–48 Cross-Sectional view of gravel-pack screen showing the attachment of the wire to the base tube.

less Steel Sprayed Sucker Rods." *Materials Performance,* 19 (1980), p. 14.

Connell, J.G. and R.M. Shapiro. "Logging Operations in Sour Gas Wells." *Symposium on Sour Gas and Crude.* Tyler, Texas: SPE of AIME, 1974.

Craig, B.D. "Sour Gas Design Considerations," *SPE Monograph,* 1992.

DeWaard, C. and D.E. Milliams. *First International Conference on Internal and External Protection of Pipes.* University of Durham, England, September 1975, BHRA Fluid Engineering, publishers.

EnDean, H.J. *Champion News,* Champion Chemical Company, Houston, Texas (July 1975).

Fowler, E.D. and A.F. Rhodes. "Checklist Can Help Specify Proper Wellhead Material." *Oil and Gas Journal* (January 24, 1977), p. 62.

Greer, J.B. "Metal Thickness and Temperature Effects in Casing and Tubing Design for Deep, Sour Wells." *47th Annual SPE Meeting,* San Antonio, Texas, 1972.

Greer, J.B. and W.E. Holland. "High Strength Heavy Wall Casing for Deep Sour Gas Wells." *55th Annual Fall Meeting of the Society of Petroleum Engineers.* Dallas, Texas, September 1980, JPT 1981.

Ikeda, A.; S. Nagata; T. Tsumura; Y. Nara; and M. Kowaka. "Development of High Strength Oil Country Tubular Goods Highly Resistant to Sulfide Stress Corrosion Cracking." Hollywood, Florida: API, 1977.

Kane, R.D. and J.B. Greer. "Sulfide Stress Cracking of High Strength Steels in Laboratory and Oilfield Environments." *Journal of Petroleum Technology* (1977), p. 1483.

Kane, R.D. and W.K. Boyd. "Materials Technology for Oil and Gas Production." *Alloys for the 80's.* AMAX, 1980, p. 225.

King, J.A. and P.S.C. Badelek. "Performance of Valve Materials in Wet H_2S and CO_2 Contaminated Hydrocarbons." *Corrosion/82,* Houston, Texas: NACE, March 1982.

Kolts, J. "Temperature Limits for Stress Corrosion Cracking of Selected Stainless Steel and Nickel-Base Alloys in Chloride-Containing Environments." *Corrosion/82,* Houston, Texas: NACE, March 1982.

Mehdizadeh, P. "Effect of Metallurgical Variables on Corrosion of Sucker Rods in Brine Containing CO_2 and H_2S." *Materials Performance,* 14 (1974), p. 13.

Metal Progress Databook. ASM. Mid-June 1981, p. 164.

Moore, K.H. "Learn to Identify and Remedy Sucker Rod Failures." *Oil and Gas Journal* (April 9, 1973), p. 73.

National Association of Corrosion Engineers (NACE). "The Field Testing of 32 Alloys in the Flow Stream of Seven Condensate Wells." *Publication No. 50-3.* July 1950.

National Association of Corrosion Engineers. *Standards MR-01-75* and *RP-04-75.*

"CO_2 Corrosion Resistance of NK Cr13 and NK Cr9." *Nippon Kokan (NKK) Technical Bulletin.* Second Edition. February 1982.

Nippon, Kokan. Unpublished Research.

Seger, F.O. and I. Maroofian. "Erosion, Cavitation and Abrasion Resistance of Choke Trim Materials," Paper No. 4710, *16th Annual OTC.* Houston, TX (1984).

Snape, E. and D. Van Rooyen. "Corrosion of Sucker Rod Steels in a Simulated Sweet Oil Environment." *Meeting of the Petroleum Division of ASME.* Dallas,

Texas, September 1968.

Steward, W.B. "Sucker Rod Failures." *Oil and Gas Journal* (April 9, 1973), p. 56.

Sumitomo Metal Industries Ltd. Unpublished Research.

Thomas, P.D. "Steels for Oilwell Casing and Tubing." *92nd AIME Annual Meeting*. Dallas, Texas, 1963.

Treseder, R.S. and T. M. Swanson. "Factors in Sulfide Corrosion Cracking of High Strength Steels." *Corrosion Journal*, 24 (1968), p. 31.

Vaughn, G.A. and Hung-Erh Chaung. "Wireline Materials for Sour Service." *Materials Performance*, 21 (1982), p. 44.

Watkins, M. and J.B. Greer. "Corrosion Testing of Highly Alloyed Materials for Deep, Sour Gas Well Environments." *Journal of Petroleum Technology* (1976), p. 698.

5

METALLURGY FOR ENHANCED OIL RECOVERY

This chapter deals with metallurgy used in processes that extract oil from reservoirs after primary and secondary (waterflooding) production have essentially depleted the easy-to-remove reserve. Christmas tree metallurgy for waterflooding as well as downhole tubulars are usually conventional and are described in Chapter Four.

Tertiary recovery involves a number of different methods for recouping oil from a reservoir and can be divided into three principal types: thermal, chemical, and gas. These categories are further broken down as shown in Table 5–1. *Steamflooding* and *CO_2 miscible floods* are the most common forms of enhanced oil recovery (EOR). Moreover, there are many variations and combinations of these EOR methods. Some are used in small pilot studies and little information is available on them. Chemical EOR methods such as polymer flooding do not represent a unique challenge to existing metallurgy, so wells are completed with conventional carbon-steel equipment. Therefore, only two methods, thermal recovery and gas flooding, will be dealt with since they represent the greatest materials challenge and a larger literature base is available for them.

Although forms of enhanced oil recovery such as firefloods have been around for almost half a century, little information is available in the literature on the metallurgy utilized and the equipment problems encountered in tertiary recovery. Therefore, this chapter is a brief review of the more common forms of EOR and some of the metallurgy used in these processes.

Fireflooding (*In Situ* Combustion)

In *fireflooding,* air is injected into a well and ignition is started in the reservoir. As combustion progresses, the combustion zone moves away from the injection well where oxygen (air) is continually supplied to the production wells. During the burn, steam and other gases from combustion displace oil and connate water ahead of the combustion front.

Adjacent to and just ahead of the combustion front, the temperature reaches a plateau on the order of 200–350°F, while in the combustion zone, temperatures in the range of 600–1200°F can occur. The combination of combustion gases, steam, and temperature produce an exceptionally severe corrosion environment composed of oxygen, organic acids, carbon dioxide, sulfides, and chlorides. Examples of various products formed from combustion of different reservoir fluids are shown in Table 5–2. A typical gas composition arriving at a production well is shown in Table 5–3.

This aggregate of corrodents is transported to the producing wells throughout the fireflood and as such offers a serious materials problem. Results of both laboratory and field-testing of materials in fireflood environments are shown in Table 5–4 and from autoclave testing in Table 5–5 at various temperatures. It is readily apparent that the commonly used oilfield material, carbon steel, is inappropriate for use in a fireflood and that even higher alloys such as 410 stainless steel (12% Cr) do not have adequate resistance to this severe environment. Furthermore, it is obvious that one alloy does not behave the same in all reser-

Table 5–1

Thermal methods:
Steam
Combustion in situ
Chemical methods:
Micellular-polymer
Polymer
Caustic
Other chemical
Gas methods:
CO_2 miscible
Hydrocarbon miscible
Other gases

(After *Oil & Gas Journal,* April 5, 1982)

Table 5–2 Analysis of Combustion-Tube Effluents from Various Reservoir Fluids.

		Reservoir			
		AA	BB	CC	DD
Water	pH	4.0-7.0	3.8-7.2	3.7-7.4	4.5-7.3
	Cl^-	65%	11%	7.2%	0.3%
	SO^-_4	0.22	0.16	0.44	0.3
	Acetic	<0.03	0.03-0.32	0.01-0.08	ND
	Formic	<0.03	ND	ND	<0.03
Oil	Acetic	<0.03	ND	<0.03	Tr
	Formic	<0.03	ND	<0.03	ND
Gas	H_2S	Tr	ND	ND	ND
	S^-	1.5	1.1	7	3.7
		Corrosion Rates at 90°C (195°F), mpy			
	Carbon steel	34.7	-	57.6	87
	Incoloy 825	<0.1	-	1.6	-
	Inconel 625	<0.1	-	1.1	-

ND = Not determined
Tr = Trace
(After Kohut et al.)

voirs since 12% Cr had a corrosion rate of 824 mpy and 14.9 mpy in the two different reservoirs described in Table 5–4.

The ignition well also presents a problem from the standpoint of a high-temperature, highly oxygenated environment. Ignition in wells has been classified as either spontaneous or artificial. Some reservoirs are susceptible to spontaneous combustion when contacted with sufficient air. In these wells, combustion occurs a short distance from the wellbore and then burns back into the borehole. This can result in temperatures up to 1,700°F and has resulted in failure of pipe in the hole by creep or melting. Even stainless-steel liners have been reported to have melted following ignition.

The most widely used artificial means of ignition is electric or gas bottom-hole heaters. This method produces temperatures in excess of 1,200°F.

A review of firefloods by Chu described well completions for injectors and producers from various companies. A reorganized presentation of this work is shown in Table 5–6.

Field experience has indicated that corrosion and erosion are the two most serious materials problems. Mild steel has been reported to have weight losses

Table 5–3 Ambient Temperature, Steady-State Composition of Gas at Production Well During Fireflooding.

Component	Concentration (%)
H_2	2.5
O_2	1.0
N_2	78.0
CO	5.0
CO_2	10.0
CH_4	2.5
H_2S	1.0

(After Johnson and Magnani)

of 73% in 2 months. Stainless steel is also susceptible to rapid failure by chloride cracking, intergranular corrosion, and pitting attack. In some fields, 316 stainless steel and titanium have performed satisfactorily and are used for tubing and liners opposite the production interval. Corrosion and erosion failures of casing, tubing, chokes, valves, and fittings have been reported at producing wells.

A limited amount of work on valve-stem materials for producing wells has shown that 316 stainless steel is not resistant enough for this application and that measures such as metallic coatings are required (see Table 5–7).

Care must be taken in selecting oxygen injection equipment since each alloy has an oxygen threshold above which once ignited, combustion is sustained. Moreover, oxygen compatible pipe dope and elastomeric seals must be selected. An aid to selecting materials for oxygen service has been published by ASTM as Standard G63 (see references).

Table 5–4 Comparison of Corrosion Rates in Field and Laboratory Environments

Alloy	Reservoir AA (60°C) Field	Reservoir AA (60°C) Laboratory	Reservoir DD (90°C) Field	Reservoir DD (90°C) Laboratory
Carbon steel	Corroded away	2640 mpy	114mpy	87 mpy
12 Cr stainless	824 mpy	960	14.9	...
Incoloy 800	0.6	1.2	0.7	...
Incoloy 825	0.2	0.1	...	...
Inconel 600	477	334	...	...
Inconel 625	...	...	<0.1	<0.1

(After Kchut et al.)

Table 5-5 Autoclave Corrosion-Test Results in Simulated Fireflood Environment D

Test Temperatures, °C (°F)	140 (280)	175 (350)	205 (400)
Carbon steel	2,900–4,000 mpy	...	...
T 410 (12% Cr)	120–400	...	...
Incoloy 600	140–300	...	...
Incoloy 800	28	...	...
Inconel 825	0.4–1.1	22–46 mpy	...
Inconel 625	...	<0.1	0.9–6.4 mpy
Hastelloy C-276	...	2.6	0.8–1.4
MP 35 N	...	3.6–4.8	14.1
Hastelloy G	...	5.5	27.4
Ti 50 A	...	...	0.7–66.1

D = 100 psi carbon dioxide, produced brine, 0.1% acetic acid, 0.1% formic acid, 0.1% sulphur dioxide.
0.2% organic compounds.
(After Kohut et al.)

Steamflooding

There are two types of steam injection: continuous and cyclic. The first type involves continuous steam injection to the reservoir while the other involves intermittent injection—often referred to as huff 'n' puff—in which steam is injected for a short time before the well is returned to production.

As in fireflooding, steam is injected into a number of wells and oil is produced from others. A steam-saturated zone forms around the injection well, while farther away is a zone containing condensed steam. Temperatures in the steam zone, which are equal to the steam temperature, decline over a distance from the injection well to a final temperature equal to the reservoir temperature. Steam temperatures can attain 500–700°F although lower-temperature steam has also been used successfully.

One of the major problem areas in steam injection wells is casing failure caused by compression, with most of the failures located at the connections. *Telescoping* and *buckling* are the most frequent forms of damage. It is believed that extensive thermal-linear expansion occurs during steam injection because the high temperatures cause excessive compressive stresses on the casing; during cooling, the stress reverses to tension and failure occurs.

Unstressed casing elongates about 8 in. per 1,000 ft for every 100°F increase

Table 5–6 Fireflood Well Completions

Field	Injectors Casing	Injectors Tubing	Injectors Liner	Producers Casing	Producers Tubing	Producers Liner
2	J-55				316 SS	316 SS
4			304 SS			304 SS
7	304 SS Shoe joint			304 SS on Bottom		
10	K-55			K-55		
13		Incoloy on bottom				
14	J-55	Inconel on bottom				
17				J-55		
18	J-55 H-40			J-55	19	H-40
24		9 Cr 1 Mo				

(After Chu,© 1981, SPE)

in temperature. If both ends of the casing are fixed such as in a conventional completion, the stress created by thermal expansion is:

$$S = \alpha (\Delta T) E$$

where:

S = stress (psi)
α = coefficient of thermal expansion (7.0×10^{-6} in./in./°F)
T = increase in temperature over the average formation temperature (°F)
E = modulus of elasticity (for steel: 30×10^{6} psi)

Since α and E are constants for steel, the following simple relationship for stress is:

$$S = 210(\Delta T)$$

As is immediately obvious from this formula, the stress is not a function of the length of the tubing or casing; rather it is solely a function of temperature change. Referring back to the earlier temperature regime of 500–700°F, using this formula, stresses on the order of 84,000–126,000 psi could be attained assuming an average reservoir temperature of 100°F. Comparing these values with the data in Table 5–8 on elevated temperature-tensile properties of seamless casing and tubing at 700°F, the yield strength would be exceeded on all grades.

Table 5-7 Operations to Failure for Letdown Valve Stems

Material	Operations
Type 316 stainless steel	30
Stellite No.6	100
Composite W/TiB_2*	187**
Composite W/TiB_2, set 2	450***

* Type 440C Stainless Steel shank with Kennametal K151A tip and with TiB2 coating
** Failure occurred in threaded end of valve handle, not at tip.
***No failure after 450 operations; test continuing.
(After Johnson and Magnani)

Figures 5–1 and 5–2 present the yield and tensile data from Table 5–8 in graph form. From these curves, P-105 or P-110 have the most desirable properties for exposure to steam injection. These enhanced properties of P grades over the other grades are attributable to the chromium and molybdenum content of the steel, which helps develop good high-temperature properties such as creep and stress rupture resistance. *Creep* is the time-dependent increase in length of a metal subjected to a constant tensile load at elevated temperatures. Creep failure is usually defined as a specific amount of dimensional change occurring over a certain time increment. Stress rupture represents complete failure (fracture) of the material by creep.

The potential for failure of fixed casing can be reduced if the following temperature limits set by Gates are observed:

Grade	Surface Temp. Range, °F
H-40	270–330
J-55	350–410
N-80	470–580
P-110	620–730

This problem can be further decreased by using a thermal packer, which often will reduce the temperature by 120°F. More frequently, the wells are designed with expansion joints to permit vertical expansion of the tubing. Also, wellheads are available that allow for casing expansion through a stuffing box and for expansion of the tubing in a tubing hanger designed for thermal expansion. Many completions use no expansion joint but instead run gravel-pack liners that have a lead or brass hanger.

Another method that alleviates casing stress caused by temperature effects has been to prestress the casing during completion. This is based on the concept

Table 5–8 Elevated Temperature-Tensile Properties of Various Grades of Seamless Casing andTubing Tensile Properties*

Grade	Condition	Outside Diameter, In.	Wall, In.	Tensile Test Temp., °F	Yield Strength, psi	Tensile Strength, psi	Elastic Limit, psi	Modulus of Elasticity, 1,000 psi	Percent Elongation, 2 in.	Percent Reduction of Area	Room Temp., Brinell Hardness
J55	Hot rolled	13⅝	0.313	RT	59.320	107,950	52,530	25,750	21.5	45.0	207
				300	55,820	100,040	50,750	24,650	14.0	38.4	212
				500	57,870	122,125	42,050	23,800	18.75	19.1	229
				700	53,130	98,490	37,925	21,550	26.25	57.5	201
				900	44,445	68,030	28,000	24,000	24.9	55.1	201
N80	Normalized	13⅝	0.500	RT	83,750	124,050	69,550	27,900	19.0	44.8	248
				300	80,150	120,675	65,950	25,400	14.0	40.6	248
				500	83,970	133,500	59,850	21,600	22.75	36.7	255
				700	74,415	116,775	51,900	16,300	24.0	61.0	248
				900	62,715	85,000	40,650	15,700	25.75	67.2	248
P105	N&T	3	0.375	RT	106,800	136,275	78,050	27,600	13.0	42.1	277
				300	102,100	131,600	73,500	26,500	11.0	46.2	269
				500	104,850	139,100	70,950	26,200	15.5	42.5	285
				700	95,900	129,100	67,500	25,200	18.25	59.7	269
				900	86,640	107,375	59,500	18,700	15.25	57.4	269
P110	Q&T	10¾	0.547	RT	128,825	145,500	120,200	27,900	14.5	44.5	302
				300	114,850	139,100	95,000	27,300	11.5	34.2	293
				500	110,025	146,050	87,750	25,500	18.0	38.6	285
				700	100,310	124,650	76,800	24,100	21.25	76.2	277
				900	78,545	92,145	51,400	19,200	21.5	77.3	277

*Average for two tests
RT = room temperature
(After Gates and Holmes)

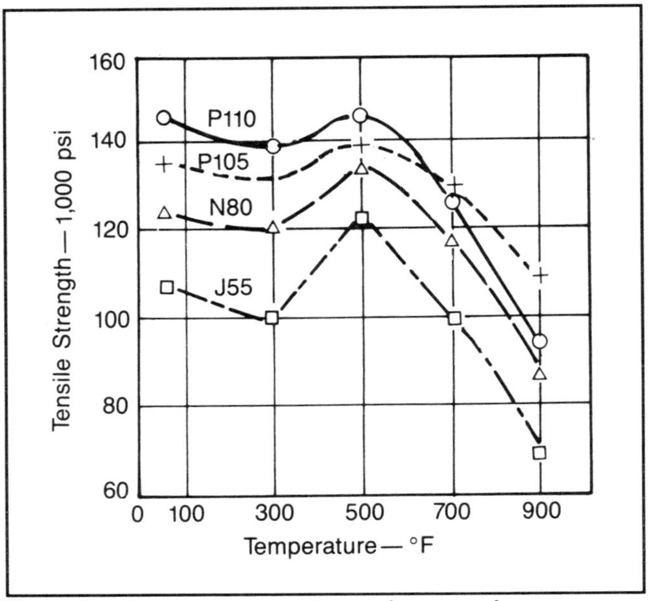

Figure 5–1 Tensile strength as a function of temperature for some common tubulars (After Gates and Holmes).

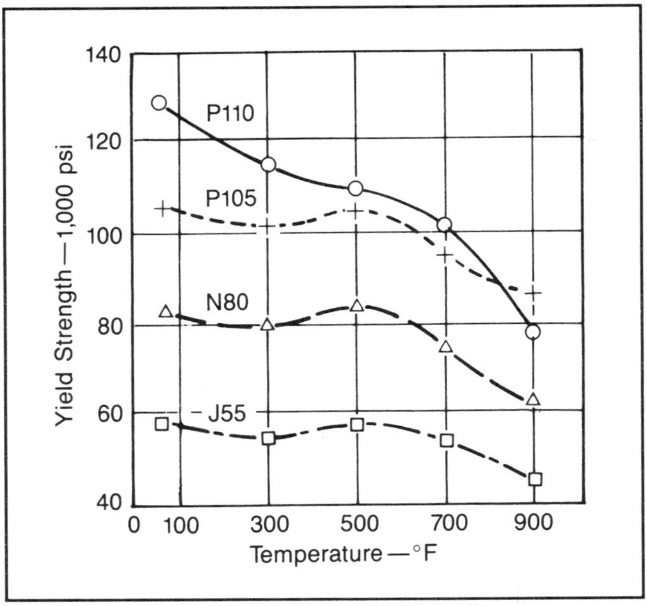

Figure 5–2 Yield strength as a function of temperature for the same alloys shown in Figure 5–1 (After Gates and Homes).

that, by placing the casing in tension before cementing, it will not exceed its yield strength in compression when steam injection commences. Since the casing must pass from tension through a neutral point of no stress to begin compressive loading, prestressing the casing offers the capability to run lower grade pipe at higher temperatures yet remain below the compressive yield strength. Once the yield strength is exceeded in compression, on cooling the joint strength may not be sufficient and failure will occur by tensile overload.

Production wells are often completed the same way as injection wells, normally using H-40 or J-55 pipe. Also, the temperature in the producing wells is much lower, reducing the thermal expansion problem. Again, wellheads that allow expansion of the tubing and casing are frequently used, just as they are used in injection wells. Production wells are susceptible to increased abrasion because of the high sand production that accompanies steamflooding. Some of this erosion can be reduced by hardfacing critical areas in valves.

Carbon Dioxide (CO_2) Flooding

Carbon dioxide flooding is performed in one of three ways: injection of pure liquid CO_2 followed by water or carbonated water; injection of carbonated water; and injection of carbon dioxide at sufficient pressure to produce miscibility with the reservoir oil.

Another method that is sometimes classed as secondary recovery is the injection of high-pressure carbon dioxide gas followed intermittently by water injection to enhance recovery. This latter method is often referred to as WAG (water-alternating-gas). The most well-documented field (unit) in which this technique has been applied is the SACROC unit in the Kelly Snyder field in Texas.

The discussion of CO_2 flooding in this section will be primarily centered on the WAG method because it presents the greatest material problem and represents the most common method of CO_2 flooding. Also, once carbon dioxide mixes with connate water, production wells experience essentially the same problems regardless of injection technique. Furthermore, on the injection side, pure, dry carbon dioxide is noncorrosive to engineering alloys in the temperature regimes of flooding. This behavior of dry carbon dioxide and the effect of impurities are presented in Figure 5–3. Here, the effects of water content, oxygen content, and velocity are described. These data are presented as weight gain instead of weight loss, but they illustrate the same effects. It is evident that flow rate and moisture content have a large effect on corrosion rate.

Corrosion of injection wells and producing wells can be quite severe during CO_2 flooding, and various steps have been taken to reduce failures of equipment

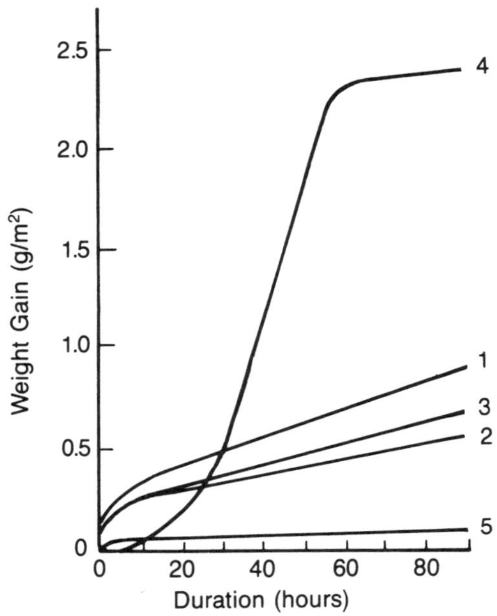

Curve 1: Gas Content (vol%)

CO_2	: 95.0%
H_2	: 3.0%
N_2	: 1.6%
O_2	: 0.3%
H_2S	: 5.3 ppm
H_2O	: Saturated at 21°C
Flow rate	: 0.5 l./min.

Curve 2: Same as 1; dried to −20°C dewpoint

Curve 3: Same as 1; 0_2 raised to 1.7%

Curve 4: Same as 1; gas flow increased to 5 l./min.; reduced to 0.5 l./min. at 60 hr

Curve 5: Gas content (vol%)

CO_2	: 96.6%
N_2	: 0.22%
O_2	: 0.22%

Methane + butane = 1.58%

Figure 5–3 Weight gain of steel in flowing gas at 70°F (After Winkler and Nemeth).

in both types of wells. Information accumulated from the literature is presented that describes the steps taken to mitigate corrosion in CO_2 flooding as well as some of the failures that led to these materials selections.

Injection trees and wellheads originally constructed of carbon or low-alloy steel components that were either plastic coated or electroless-nickel plated have failed quickly under WAG conditions as a result of damage to the coating or plating, which presented areas for highly localized pitting and galvanic attack. These trees were replaced with 316 stainless-steel Christmas trees in which the gate valves had 316 stainless-steel gates or ceramic gates. Also, aluminum-bronze valves with ceramic gates appear to be a successful alternative, according to Gill.

In other cases, 410 stainless-steel trees and valves were severely pitted under scales formed from the injection water but Type 316 replacement components showed no apparent problem. Even though 316 stainless steel is susceptible to chloride SCC, the temperatures of these wellheads is low enough (150°F) to avoid this type of failure. However, 316 stainless steel welded fittings have suffered intergranular corrosion in the weld and heat-affected zone. Stainless steels are also quite susceptible to crevice corrosion in areas such as flanges where gaskets can absorb moisture. This has been essentially solved using nonabsorbing gaskets.

Tubing hangers are usually made of 316 stainless steel, and tubing is restricted to J-55 grade steel coated with plastic. However, the coating is not always reliable and CRA tubing is sometimes used in WAG system. Of course, when no water is present, bare carbon steel will suffice.

Christmas tree valves on producing wells with electroless-nickel-plated gates have displayed numerous failures at coating breaks requiring application of 17-4 PH stainless steel gates. Full-length normalizing of J-55 is required to avoid ringworm corrosion. In some fields the most effective production tubing is 13 Cr.

Rod-pump wells have encountered serious corrosion of the rods. Pumps are run with Monel 400 for barrels, plungers, and all wetted parts and have performed without problems. Also, some pump barrels made of Monel 400 are spray metal coated with 316 stainless steel. Rod strings in some instances have only lasted one month, requiring replacement of the entire string. New steel rods plasma coated with 316 stainless steel, then coated with a baked-on epoxy, have performed satisfactorily.

Wells with submersible pumps have also experienced problems caused by corrosion. Most failures were caused by attack of the pumps and motor housing. Pitting was the major form of corrosion aggravated by iron sulfide and mineral scales. Electroless-nickel coating of the pump and motor housing usually results in premature failure caused by mechanical damage of the coating. Again, the unfavorable galvanic situation of a large nickel cathode and small steel anode lead to rapid attack. Some success has been achieved by plasma spraying 316 stainless steel and Monel onto these parts followed by overcoating them with epoxy. However, because of the potential for corrosion at damaged areas of this coating, the consideration of manufacturing the pump and motor housings out of 300 series stainless steel is a better alternative. Pump internals made of Ni-resist cast iron have not had appreciable trouble.

Although stress-corrosion cracking strictly caused by carbon dioxide has not been reported from field data, it has been observed in the laboratory and should be considered in the design and application of equipment. N–80 tubing subjected to high partial pressures (300 psia) of carbon dioxide and deformation beyond the yield point was found to stress-corrosion crack at ambient temperature (see Table 5–9). One explanation by the investigators stated this stress-corrosion cracking was attributed to dissolution of the sulfide inclusions in the steel, which led to hydrogen sulfide evolution and eventually to sulfide-stress cracking.

Much of the data developed for materials in CO_2 gas systems discussed in Chapter Four is also applicable to WAG systems. However, since the temperatures and partial pressure of CO_2 are not high other materials not acceptable in Chapter Four can be for WAG. Thus when H_2S is not present in quantities greater than 10 ppm copper base alloys can be quite effective.

Inert Gas Injection

Inert gas injection, a form of miscible gas flooding, involves the injection of an inert gas, usually nitrogen, into the reservoir under high pressure. A typical method for procuring this inert gas is passing exhaust gas from compression engines over a catalyst bed. The composition of such a gas is shown in Table 5–10. This type of process is also commonly used for gas-pressure maintenance of reservoirs in which inert gas is injected into the gas cap to maintain the reservoir pressure, extending the life of the primary production.

At first glance, the gas composition does not look too aggressive, and one would expect only some CO_2 corrosion when water is present. However, field experience has shown that stress-corrosion cracking can occur in the presence of CO/CO_2 gas mixtures. Field failures of J-55 and N-80 tubing resulted when injection gas leaked through the tubing connection and contacted the packer fluid. Cracking frequently initiated at small pits on the outside surface of the tubing and progressed in a transgranular fashion. Although the mechanism for such cracking has not been determined, a laboratory study of this type of cracking led to the following conclusions by Kowaka and Nagata:

1. Stress-corrosion cracking did not occur in water containing only carbon monoxide or carbon dioxide.

2. Cracking was transgranular and only occurred when water was present.
3. Cracking occurred over a wide range of CO/CO_2 ratios.
4. Susceptibility to cracking decreased with increasing temperature.
5. No cracking occurred in 18 Cr-10 Ni stainless steel.
6. Cracking was apparently caused by stress-corrosion cracking rather than hydrogen embrittlement.

Table 5-9 Stress-Corrosion Cracking Tests in CO_2 Solutions in Comparison with Other "Nonsulfide" Media

Environment	Material	Hardness	% Deformation	Failure Time
15 psia air[1]	N-80-2	35	105,90,70,60,40,20	NF 211 days[3]
315 psai CO_2[1]	N-80-1	38	130,100,80,60,40,20	NF 695 days[3]
300 psia CO_2[1]	N-80-2	18	130	NF 520 days[3]
300psia CO_2[1]	N-80-2	25	130	NF 520 days[3]
300 psia CO_2[1]	N-80-2	34	110	326 days[3]
300 psia CO_2[1]	N-80-2	33	115	22 hours
300 psia CO_2[1]	N-80-2	33	115	NF 84 days[3]
300 psia CO_2[1]	N-80-2	33	100.80	NF 112 days[3]
0.013 N HCl	N-80-2	34	110	5.8 hours
1.07 N HCl	N-80-2	34	110	0.35 hour
11.6 N HCl	N-80-1	36	100	0.03 hour[2]
11.6 N HCl	N-80-1	36	80	0.25 hour
9.4 N H_2SO_4	N-80-1	39	100	0.18 hour

[1]Gases bubbled through 5% NaCl
[2]Triplicates
[3]Failure
(After Hudgins et al.)

Table 5-10 Typical Inert Gas Composition

Component	Mole %
Hydrogen	2.3
Methane	0.067
Ethane	0.008
Water	0.001
Nitrogen	83.8
Oxygen	0.029
Argon	0.96
Carbon dioxide	9.4
Carbon Monoxide	3.4

(After Bowman et al.)

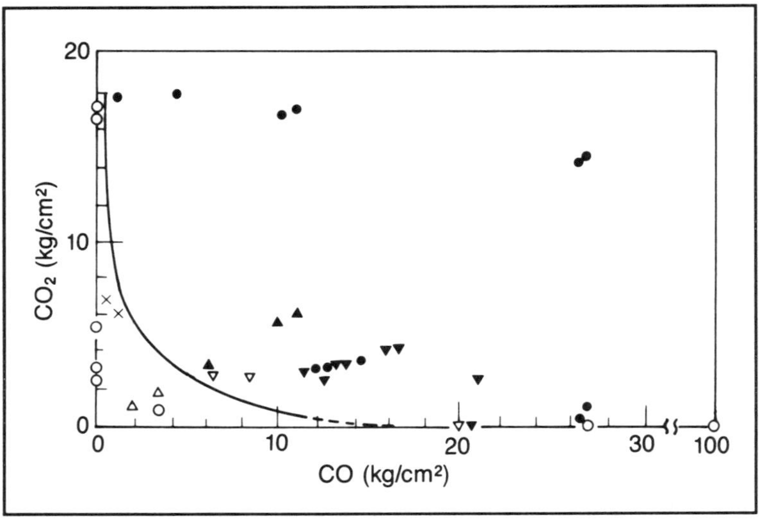

Figure 5–4 Effect of partial pressure of CO and CO_2 on stress-corrosion cracking of carbon and manganese steels between 18-70°C for 168–1,000 Hours. Closed symbols and X denote cracking and open symbols no cracking. (After Kowaka and Nagata).

The last conclusion is not completely in accord with the previous five conclusions since decreasing susceptibility with increasing temperature is more characteristic of hydrogen-stress cracking (HSC) than stress-corrosion cracking. In addition, carbon monoxide is also an effective hydrogen recombination poison and as such could promote the ingress of hydrogen into the steel. Thus, this form of cracking actually may be HSC but elucidation of the mechanism must await further research.

Figure 5–4 shows the reason for Conclusion Three and the need for a better understanding of this form of cracking since, in thermal recovery methods, as well as inert gas injection systems, various concentrations of carbon monoxide and carbon dioxide are developed and pose a potential cracking problem.

Describing the various problems associated with enhanced oil recovery presents a new challenge to the corrosion/materials engineer in the application of materials. Many of these environments are similar to basic petroleum production (e.g., hydrogen sulfide, carbon dioxide, chlorides, etc.), but some are not, such as significantly higher temperatures, oxidizing environments, increased sand production with associated erosion, and new cracking phenomena such as CO_2/CO environments. All represent new areas of concern and challenge. Although there are alloys currently available that circumvent many of these problems using the right combination to avoid other associated problems such as crevice corrosion, SCC, and galvanic attack and the need for economic considerations make material decisions complex.

Experience with alloys in deep, aggressive wells should provide an information base that can be extended to the EOR realm and vice versa.

References

ASTM G63-80 "Standard Guide for Evaluating Materials for Oxygen Service" (1980).

Bowman, R.W.; A.K. Dunlop; and J.P. Tralmer. "CO/CO_2 Cracking in Inert Gas-Miscible Flooding." *Materials Performance*, 16 (1977), p. 28.

Chu, Chieh. "State-of-the-Art Review of Fireflood Field Projects." *Second Joint SPE/DOE Symposium on Enhanced Oil Recovery.* Tulsa, Oklahoma, April 1981.

Clark, G.A.; R.G. Jones; W.L. Kinney; R.E. Schilson; H. Surkalo; and R.S. Wilson. "The Fry in Situ Combustion Test—Field Operations." *Journal of Petroleum Technology (1965), p. 343.*

Faroug, Ali, S.M., and R.F. Meldau. "Current Steamflood Technology." *Journal of Petroleum Technology 31* (1979), p. 1332.

Gates, C.F. and B.G. Holmes. "Thermal Well Completions and Operation." *Proceedings of the Seventh World Petroleum Conference.* Volume 3, Mexico, 1967, p. 419.

Gill, T.E. "Ten Years of Handling CO_2 for Sacroc Unit." Annual SPE of AIME meeting. New Orleans, September 1982.

Hudgins, C.M.; R.L. McGlasson; P. Mehdizadeh; and W.M. Rosborough. "Hydrogen Sulfide Cracking of Carbon and Alloy Steels." *Corrosion* 22 (1966), p. 238.

Johnson, D.R. and N.J. Magnani. "Materials Studies for Thermal, Enhanced Oil Recovery Processes." *Proceedings of the Fourth Annual DOE Symposium on Enhanced Oil and Gas Recovery.* Tulsa, Oklahoma, 1978.

Kohut, G.B.; J.R. Caldwell; and R.A. Morris. "Fireflooding—A Materials Challenge for the Eighties." *Alloys for the 80's.* Ann Arbor, Michigan: AMAX Conference, 1980.

Kowaka, M. and S. Nagata. "Stress Corrosion Cracking of Mild and Low Alloy Steels in CO-CO_2-H_2O Environments." *Corrosion*, 32 (1976), p. 395.

Moss, J.T. "Ignition Equipment and Procedures for Thermal Recovery Projects." *Twelfth Annual Southwest Petroleum Short Course.* Lubbock, Texas, 1965.

Patterson, K.W. "Downhole Corrosion Encountered in the CO_2 Flood at the SACROC Unit." *Petroleum Engineer International, 51 (1979), p. 36.*

Schremp, F.W. and G. R. Roberson. "Effect of Supercritical CO_2 on Materials of Construction." *48th Annual Fall Meeting of SPE.* Paper 4667. Las Vegas, Nevada, 1973.

Secondary Tertiary Oil Recovery Process. Interstate Oil Compact Commission. Oklahoma City, OK, 1974.

Willhite, G.P. and W.K. Dietrich. "Design Criteria for Completion of Steam Injection Wells." *Journal of Petroleum Technology (1967),* p. 15.

Winkler, G. and M. Nemeth. "The Effect of Contaminations in Carbon Dioxide on the Corrosion Caused by the Gas." *Corrosion Week, 41st Manifestation European Federation of Corrosion,* 1970, p. 788.

Zawlerucha, R., Drnevich, R.F., McIlroy, K. and Knecht, P. "Material Compatibility and Systems Considerations in Thermal EOR Environments Containing High Pressure Oxygen." *Journal of Petroleum Technology,* p. 1477 (1988).

6

SURFACE FACILITIES

Surface facilities represent a large group of equipment that is required for handling, separating, and transporting oil, gas, and water. In this chapter only those major components on the surface, downstream of the choke but upstream of a gas plant or refinery, are discussed. The major components are line pipe, vessels, tanks, rotating equipment (compressors and pumps), valves, and heat exchangers. Each of these components will be addressed in a general manner but it must be recognized that the conditions under which they operate can be quite broad.

Line pipe

Generally line pipe and piping are the same component except the former are longer sections and the latter are shorter pieces between for instance, vessels and pumps. The most common material for piping and line pipe is Grade B pipe made to either API 5L or ASTM A 106. For ease of discussion, this section will deal with API line pipe but many of the same attributes apply regardless of application. Pipe for flowlines, gathering lines and pipelines is commonly manufactured to API Specification 5L. The chemical composition and mechanical property requirements are presented in Tables 6–1 and 6–2. Similar to casing and tubing, API line pipe requirements are quite general and allow for consider-

Table 6–1 Chemical Requirements for Heat Analyses of API Line Pipe, weight percent

1	2	3	4		5		6	7	8	9
Type of Pipe	**Grade and Class**	**Carbon Max.[2]**	**Manganese Min.**	**Manganese Max.[2]**	**Phosphorus Min.**	**Phosphorus Max.**	**Sulfur Max.**	**Cb Min.**	**V Min.**	**Ti Min.**
Seamless										
Non-expanded or cold expanded	A25.Cl I	21	.30	.60	...	0.030	0.030	...	...	...
"	A25.Cl II[1]	.21	.30	.60	.045	0.080	0.030	...	...	...
"	A	.22	...	.90	...	0.030	0.030	...	...	...
"	B[7]	.27	...	1.15	...	0.030	0.030	...	...	...
Non-expanded	X42[7]	.29	...	1.25	...	0.030	0.030	...	...	...
"	X46[7], X52[7]	.31	...	1.35	...	0.030	0.030	...	...	...
Cold expanded	X42[7], X46[7], X52	.29[3]	...	1.25	...	0.030	0.030	...	...	...
Non-expanded or cold expanded	X56[4], X60[4]	.26	...	1.35	...	0.030	0.030	.005[5]	.005[5]	.005[5]
"	X65, X70, X80	(By Agreement)								
Welded										
Electric-welded or Continuous welded only	A25.Cl I	.21	.30	.60	...	0.030	0.030			
"	A25.Cl II[2]	.21	.30	.60	.045	0.080	0.030			
Non-expanded or cold expanded	A	.21	...	.90	...	0.030	0.030			
"	B[7]	.26	...	1.15	...	0.030	0.030			
"	X42[7]	.28	...	1.25	...	0.030	0.030			
Non-expanded	X46[7], X52[7]	.30	...	1.35	...	0.030	0.030			
Cold expanded	X46[7], X52[7]	.28	...	1.25	...	0.030	0.030			
Non-expanded or cold expanded	X56[4], X60[4]	.26	...	1.35	...	0.030	0.030			

"	X65[6]	.26	...	1.40	...	0.030	0.030
"	X70[4]	.23[6]	...	1.60[8]	...	0.030	0.030
Non-expanded or Cold expanded	X80[4]	0.18[8,9]	...	1.80[8,9]		0.030[9]	0.018[9]

[1]Class II Steel is rephosphorized (see Par. 1.1 for note on bending and threading properties).

[2]In Grades X42 through X65, for each reduction of 0.01 percent below the maximum carbon content, an increase of 0.05 percent manganese above the specified maximum is permissible, up to a maximum of 1.45 percent for Grades X52 and lower and up to a maximum of 1.60 percent for Grades higher than X52.

[3]For cold expanded seamless pipe in Sizes 20 in. O.D. and larger, the maximum carbon content shall be 0.28 percent.

[4]Other chemical analyses may be furnished by agreement between purchaser and manufacturer.

[5]Either columbium, vanadium or titanium, or a combination thereof, may be used at the discretion of the manufacturer.

[6]For Grades X65 in Sizes 16 in. and larger with wall thicknesses .500 in. *(12.7 mm)* and less, the chemical composition shall be as shown or as agreed upon between the purchaser and manufacturer. For other sizes and wall thicknesses, the chemical composition shall be as agreed upon between purchaser and manufacturer.

[7]Columbium, titanium, vanadium, or a combination thereof, may be used by agreement between purchaser and manufacturer.

[8]Manganese level may be increased by 0.05 percent for each 0.01 percent decrease in carbon up to a max manganese level of 2.00 percent.

[9]Levels are product analysis limits eliminating the need for check tolerances.

(courtesy API)

Table 6–2 Tensile Requirements for API Line Pipe

1	2		3		4		5
Grade	Yield Strength, Min.		Ultimate Tensile Strength, Min.		Ultimate Tensile Strength, Max.		Elongation, Min. Percent in 2 in.
	PSI	MPa	PSI	MPa	PSI	MPa	*(50.80 mm)*[1]
A25	25,000	*(172)*	45,000	*(310)*			
A	30,000	*(207)*	48,000	*(331)*			
B	35,000	*(241)*	60,000	*(413)*			
X42	42,000	*(289)*	60,000	*(413)*			
X46	46,000	*(317)*	63,000	*(434)*			
X52	52,000	*(358)*	66,000	*(455)*			
X56	56,000	*(386)*	71,000	*(489)*			
X60	60,000	*(413)*	75,000	*(517)*			
X65	65,000	*(448)*	77,000	*(530)*			
X70	70,000	*(482)*	82,000	*(565)*			
X80	80,000	*(551)*	90,000	*(620)*	120,000	*(827)*	

[1]The minimum elongation in 2 in. *(50.80 mm)* shall be that determined by the following formula:

English Formula

$$e=625{,}000\,\frac{A^{0.2}}{U^{0.9}}$$

Metric Formula

$$e=1942.57\,\frac{A^{0.2}}{U^{0.9}}$$

Where: e = minimum elongation in 2 in. *(50.80 mm)* in percent to nearest 1/2 percent.
A = cross-sectional area of the tensile test specimen in sq. in. *(mm)* based on specified outside diameter or nominal specimen width and specified wall thickness rounded to the nearest 0.01 sq. in. *(6.5 mm^2)* or .75 sq. in. *(484 mm^2)*, whichever is smaller.
U = specified minimum ultimate tensile strength, psi, (MPa).

See Appendix C for minimum elongation values for various size tensile specimens and grades. The minimum elongations for both round bar tensile specimens (0.350 in. *(8.9 mm)* diameter with 1.4 in. *(35 mm)* gage length, and the 0.500 in. *(12.5 mm)* diameter with 2.00 in. *(50.8 mm)* gage length) shall be that shown in the Area A line of .20 sq. in. in the Elongation Table of Appendix C. (courtesy API)

able variation in chemistry and strength. Generally, the lower strength grades are C-Mn steels but more sophistication is introduced both in chemistry and processing as the strength level increases, especially for Grades X60 and higher. In these higher strength grades, the carbon content is reduced considerably and small amounts of molybdenum, niobium, titanium, and vanadium may be added. These elements are added in such small quantities that this practice is referred to as microalloying. Through microalloying and control of the rolling practice, high strength line pipe steels can be manufactured with very low carbon content, Figure 6–1. As will be discussed later, the lower carbon content increases weldability and fracture toughness.

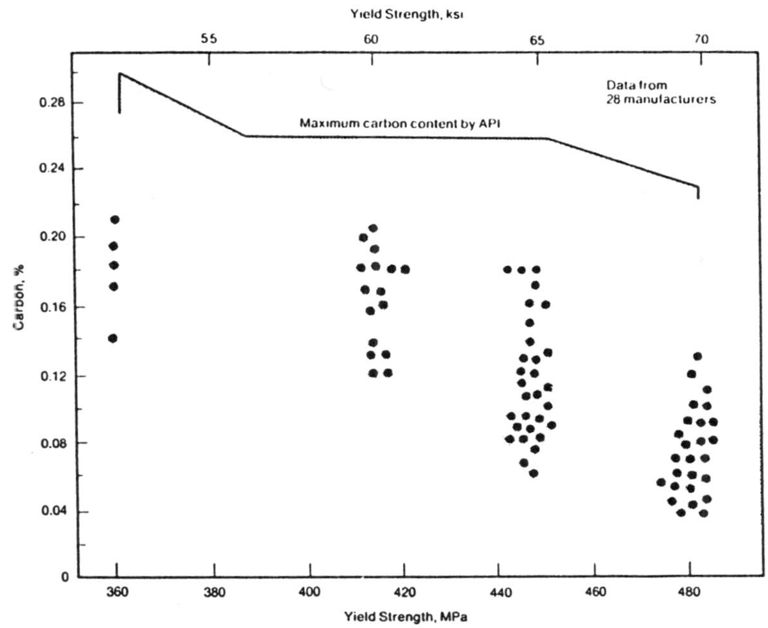

Figure 6–1 Trend in declining carbon content of linepipe steels with increasing strength since 1965 (courtesy American Welding Society).

Manufacturing Processes of Line Pipe

Since the manufacturing of ERW and seamless pipe has already been discussed in Chapter Four only the manufacturing of submerged arc pipe will be presented here. Spiral weld pipe is another process but it is not as common as the other three processes and therefore is not described.

Submerged-Arc Welded Line Pipe

The manufacturing of double submerged-arc welded (DSAW) line pipe is usually restricted to diameters of 12-¾-inch OD and above. Most DSAW pipe is manufactured from plate, although a few mills utilize strip or sheet product. Whether it is plate or strip, the first operation in the production sequence is the shearing and planing of the edges to ensure that each is parallel with the opposite side and square with the ends.

Crimping of the plate edges follows the planing operation and this is done to avoid a flat surface near the longitudinal weld seam. The crimped plate is "U"-ed by a large die with a 2,000-ton capacity or more. As the die moves down onto the plate, it forces it between sets of dies which assist in shaping the "U." The shaped plate is positioned in a bottom die, and a top die is operated by a hydraulic press which is forced down. The plate is deformed until it is almost a closed cylinder or "can."

The formed cylinder is welded longitudinally by a submerged-arc process using consumable electrodes and granular fusible flux. Usually a two- or three-wire tandem electrode technique is used and the can is welded from the inside first. Specially designed welding heads deposit the flux along the weld groove and feed the consumable electrodes from a coil mounted on a reel. Both the coil and flux holder are mounted on a boom that travels through the pipe, which is held stationary by a set of clamps. A run-on tab is attached to both ends of the can to facilitate striking the arc and setting the welding parameters. After the inside weld bead is deposited, the weld is inspected and small weld repairs are made. External welding is then performed with a welding head similar to the internal unit. The weld is thoroughly inspected after the completion of the external welding and the loose flux is removed. Defects are usually repaired using a gas shielded metal-arc, a semi- automatic submerged-arc, or low hydrogen shielded metal-arc process.

The final diameter is obtained by cold expanding the pipe either hydraulically or mechanically. Hydraulic expansion uses a retaining die around the outside whereas an internal mandrel is employed to expand the pipe mechanically. Expansion by either method not only rounds out and straightens the pipe but also severely tests the weld and increases the yield strength of the base metal. Hydrostatic testing of the pipe generally follows the cold expansion process.

During the manufacturing of the pipe, the weld seam is nondestructively inspected by either radiography, ultrasonics or a combination of both.

Defects associated with DSAW pipe include cracks in the toe of the weld, lack of cross penetration between the external and internal weld passes, shrinkage cracks in the center of the weld, weld metal porosity, and external and internal undercut in the heat-affected zone.

Line pipe weldability

It has been found that lowering the carbon content of the steel and microalloying with elements like Mo, Ti, V and Nb, coupled with processing, can significantly increase strength while also enhancing fracture resistance. Moreover, lowering the carbon and manganese content significantly improves weldability.

Good weldability is important to the successful joining of line pipe and its long-term serviceability. There are several formulas used for weldability; however, one of the most common is the IIW (International Institute of Welding) carbon equivalent (CE):

$$CE = C + \frac{Mn}{6} + \frac{(Cr + Mo + V)}{5} + \frac{(Ni + Cu)}{15}$$

where the elements are in weight percent.

The CE is an indication of a steel's potential for weld cracking due to hydrogen underbead cracking. Hydrogen enters the weld from various sources such as the electrode coating, moisture, grease, etc. The lower the CE the less chance for cracking of the weldment. Generally the CE is considered a function of the hardenability of the steel and thus reflects the final hardness, which is also important for example in sour service.

Another method for evaluating cold (hydrogen) cracking potential of a steel is with the P_{cm} equation:

$$P_{cm} = C + \frac{Si}{30} + \frac{Mn + Cu + Cr}{20} + \frac{Ni}{60} + \frac{Mo}{15} + \frac{V}{10} + 5B$$

where the elements are in weight percent. Lower P_{cm} also increases toughness as shown in Figure 6–2. The P_{cm} is only strictly valid for low carbon steels and the CE for higher carbon steels. For improved weldability the typical limits for CE and P_{cm} are 0.40 and 0.18, respectively. Besides composition, the particular method of welding is also important to the final properties of the weld. The primary welding processes used for pipe welding are shielded metal-arc (stick electrode), gas metal-arc welding (GMAW or MIG), and gas tungsten-arc welding (GTAW or TIG). Welding high carbon-equivalent base metal can lead to extremely hard welds that are susceptible to fracture in H_2S. Besides the susceptibility of the base metal to corrosion and cracking in H_2S, the welds must also be considered. Pipeline steels may suffer damage from H_2S in three ways: corrosion which was discussed in the section on H_2S corrosion, hydrogen-induced stepwise cracking (HIC), and SSC. Since corrosion by H_2S has already been dealt with, it will not be covered again.

Generally, the lower strength pipeline steels, Grades X60 and lower, do not display SSC in the base metal because of their lower hardness, but the welds in these steels may fail by SSC. As higher strength grades have been developed, the tendency to SSC of the base metal as well as the welds has increased. Figure 6–3 illustrates the changes in hydrogen damage mechanism as a function

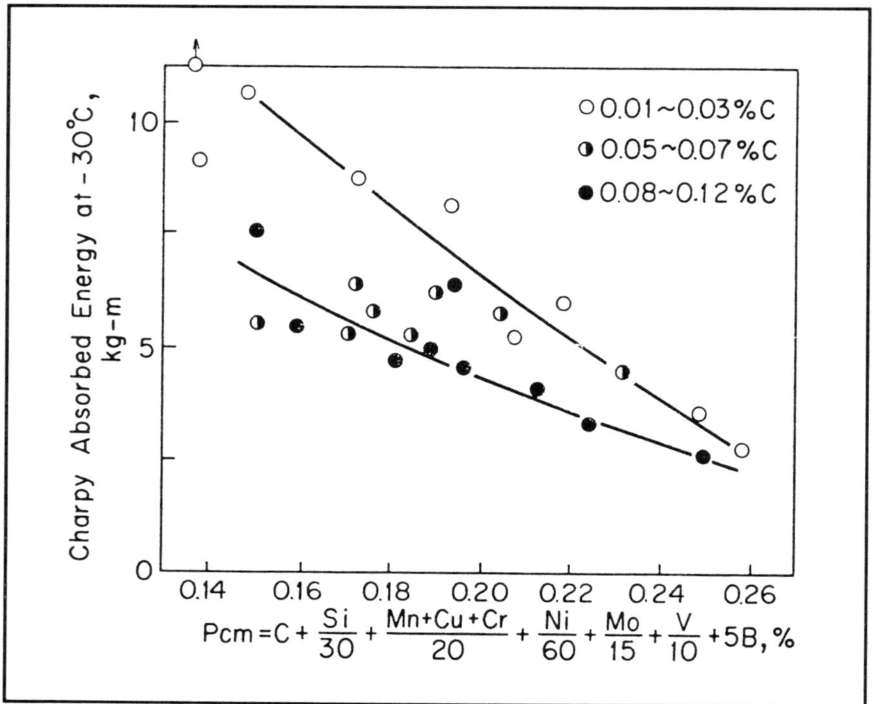

Figure 6–2 Charpy impact toughness of weld heat-affected zone as a function of the P_{cm} of the steel (courtesy Sumitomo Metal Industries).

of yield strength. As strength increases, lower hydrogen contents are necessary to cause cracking.

Even when the line pipe steel is low strength and the welds are made to provide a SSC resistant structure, another form of hydrogen damage may occur, referred to variously as hydrogen-induced cracking (HIC), stepwise cracking (SWC), blister cracking or any number of other terms. Table 6–3 shows examples of pipeline failures in H_2S. Unlike SSC, SWC typically occurs in the absence of stress and initiates at elongated nonmetallic inclusions (often MnS). Cracks linking the blisters formed at inclusions follow a stairstep pattern thus the term stepwise cracking. This feature is shown schematically in Figure 6-4 and compared to the crack path observed for SSC. For soft steels in which a high stress is applied, the fracture appears intermediate between SWC and SSC, (Figure 6–4) and is referred to by some investigators as stress-oriented hydrogen-induced cracking (SOHIC). This is by no means a universally accepted term and is only presented for distinction purposes. The mechanism of SWC is illus-

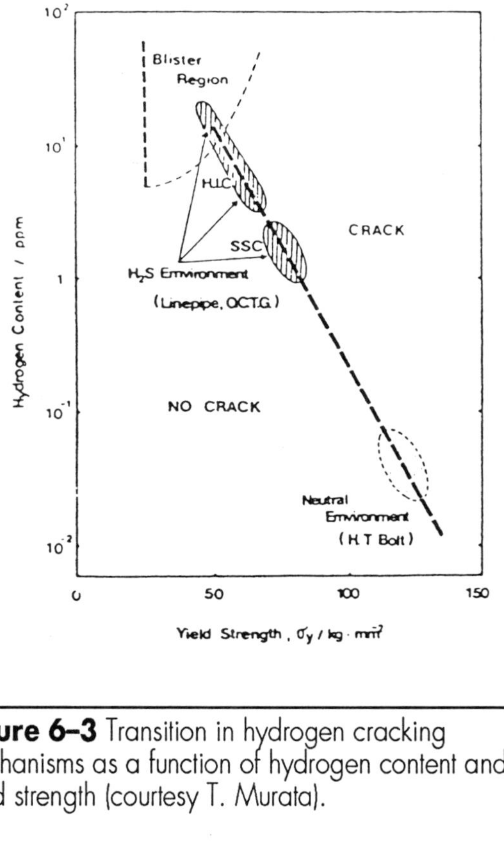

Figure 6–3 Transition in hydrogen cracking mechanisms as a function of hydrogen content and yield strength (courtesy T. Murata).

Table 6–3 Examples of Pipeline Failure Caused by Cracking Under H_2S Environment

Year	Location	Plant	Material	Environment
1954	USA	Barkerdom Gathering Pipeline	X-52 SAW 24"Ø x 0.271"t	Natural Gas 15% CO_2, 1% H_2S
1961	FRG	Gas Pipeline	X-42 ERW Annealed	$H_2S \leq 0.95\%$ $CO_2 = 8.7\%$
1972	Arabian Gulf	Offshore Pipeline (Crude Oil)	X-65 SAW	Crude Oil + H_2S (with Sea Water)
1974	Arabia	Onland Pipeline (Sour Gas)	X-42 SAW (Spiral)	Natural Gas (CH_4) 3.4% H_2S, 8.8% CO_2

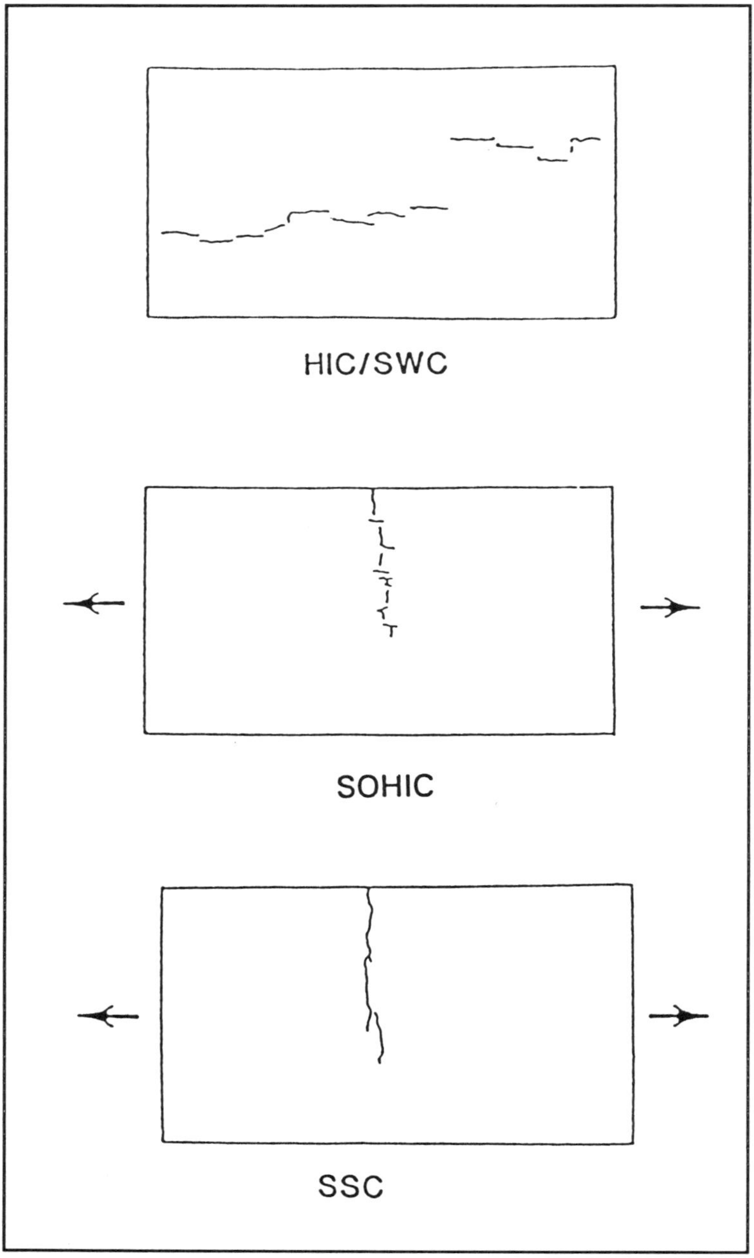

Figure 6–4 Crack morphology for HIC/SWC, SSC, and SOHIC. The arrows indicate the direction of applied stress.

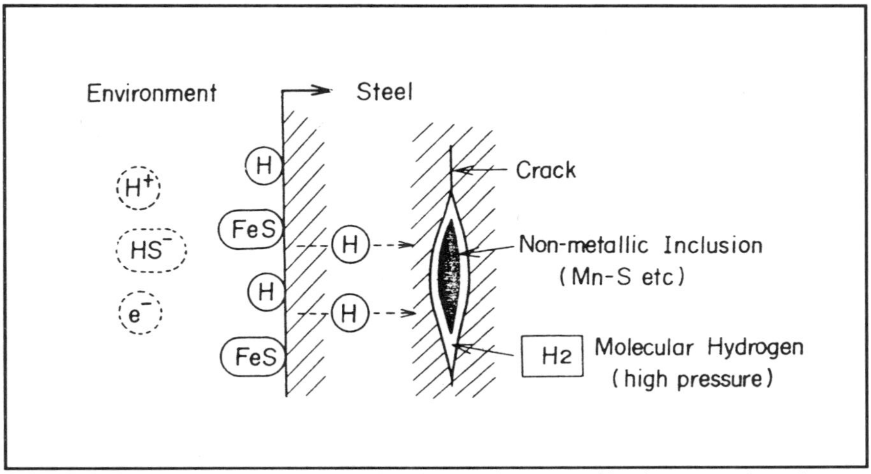

Figure 6–5 Schematic drawing for the mechanism of hydrogen-induced cracking (HIC) (courtesy Sumitomo Metal Industries).

trated in Figure 6–5. SWC can occur in the base metal or welds of pipe as can SOHIC. The term SWC is preferable to HIC since all of the numerous hydrogen damage modes are hydrogen induced. Thus, the term HIC does not distinguish one mechanism of hydrogen damage from another. However, popular opinion has favored HIC.

As with SSC, there are a number of factors that affect SWC. The primary factors are:

- Solution pH
- Hydrogen content of steel
- Temperature
- Steel chemistry
- Processing/Microstructure

Figure 6–6 shows the degree of cracking as a function of hydrogen content. After a certain threshold is reached, cracking becomes significant. The concentration of hydrogen in a metal is a function of the solution pH and the partial pressure of H_2S. Figure 6–7 shows the partial pressure necessary for HIC and the dependence on composition of the steel. Additions of Cu have been found to decrease both the corrosion rate in H_2S and the tendency to SWC.

Steel chemistry has been found to have a profound effect on SWC resistance. As might be expected, a reduction in sulfur content should increase resistance by reducing the number of nonmetallic inclusions from which SWC could initiate. This is observed to be the case in Figure 6–8. Furthermore, control of the

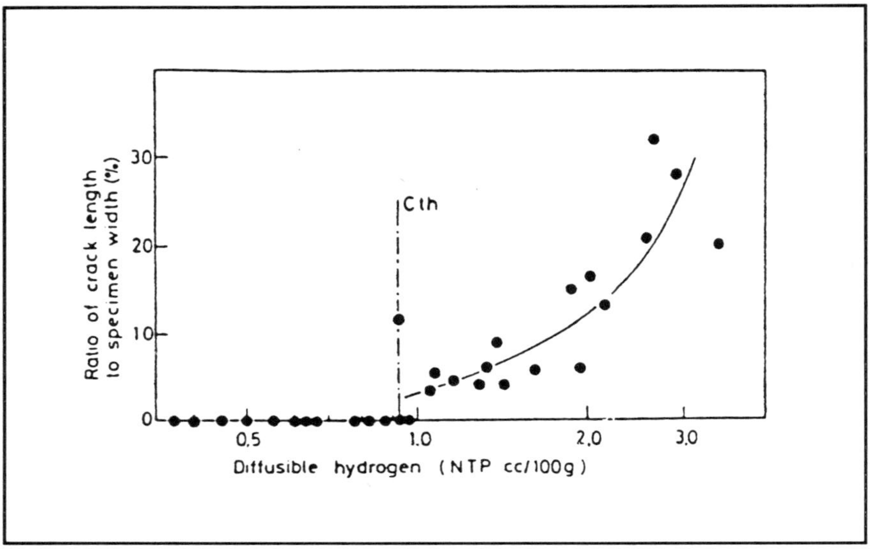

Figure 6–6 Relationship between HIC and the threshold hydrogen content (C_{th}) (courtesy A. Ikeda and NACE).

sulfide shape using Ca treatment has also proven beneficial, Figure 6–9.

Microstructure is also quite important in resisting SWC. A heavily banded (segregated) structure can reduce resistance to SWC even if all other precautions are taken. Since banding is dependent on Mn content, Mn levels are maintained as low as possible to reduce the incidence of banding. In a similar fashion for SSC, the temperature of greatest susceptibility for HIC appears to be about 30°C.

A standardized test procedure for susceptibility of pipe steels to SWC has been developed by NACE and designated TM-02-84. This procedure specifies the removal of test coupons from the pipe and the exposure of these coupons to a seawater solution saturated with H_2S. The test is run for 96 hours after which the specimen is cleaned, sectioned and metallographically prepared. The extent of cracking is measured and the results are reported as crack length ratio (CLR), crack sensitivity ratio (CSR) and crack thickness ratio (CTR). However, this NACE test method does not provide accept/reject criteria, which must be selected by the user.

It is important when considering new construction of flowlines, gathering lines, and transmission lines that HIC be considered and pipe ordered that is resistant to this form of attack. Moreover, it has been found that seamless pipe does not generally display HIC because the inclusions are not elongated; there-

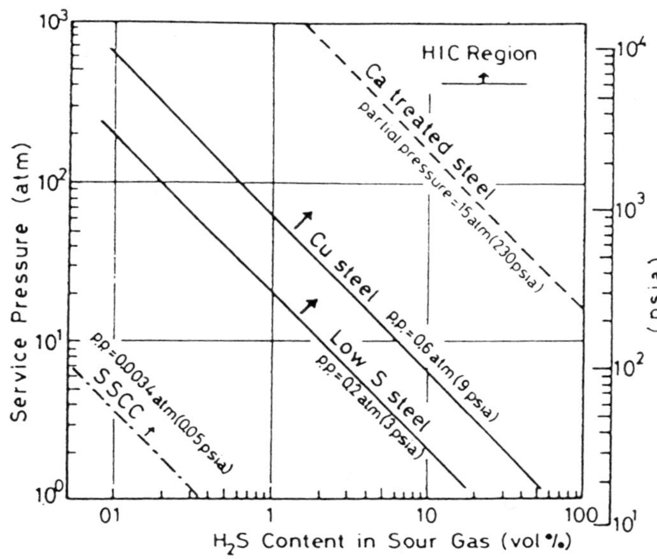

Figure 6–7 Critical H_2S partial pressure required to initiate HIC (courtesy N. Totsuka and NACE).

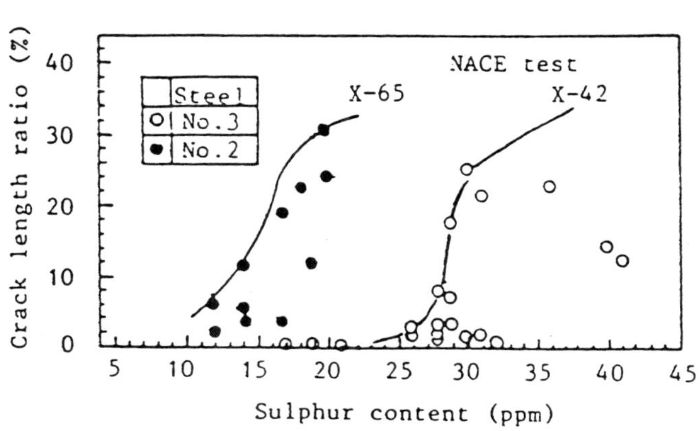

Figure 6–8 Influence of sulfur content in steel on HIC behavior measured as crack length ratio (CLR) (courtesy ASM International).

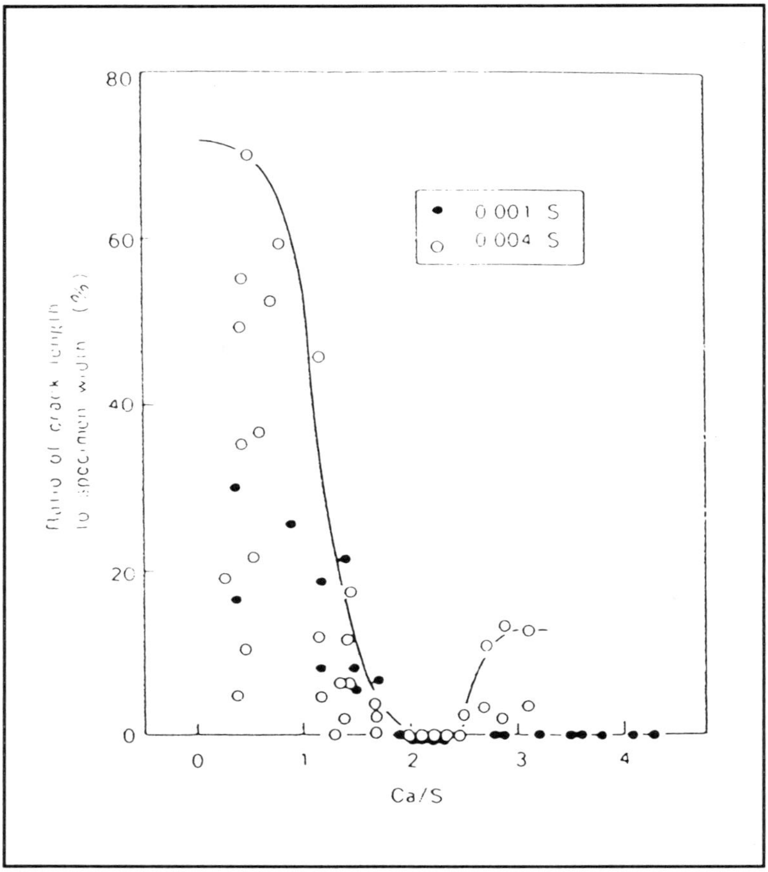

Figure 6–9 Effect of Ca/S ratio on HIC resistance in Ca-treated steel (courtesy A. Ikeda and NACE).

fore, ordering HIC-resistant seamless pipe is unnecessary. However, ERW and DSAW pipe may display HIC and precautions should be taken.

As pipeline steel strength increases to X80 and greater, the potential for SSC increases. The behavior of high strength pipeline steels to SSC is identical to other steels. Cracking in H_2S is also a function of the welding process and microstructure, Figure 6–10. The presence of increasing amounts of bainite and martensite in the weld increase the incidence of SSC.

One of the serious concerns now facing the industry is the sulfide cracking of welds. Figure 6–11 shows the relation between HAZ hardness and H_2S concentration to resist fracture. Although hard welds are expected to fail in H_2S from

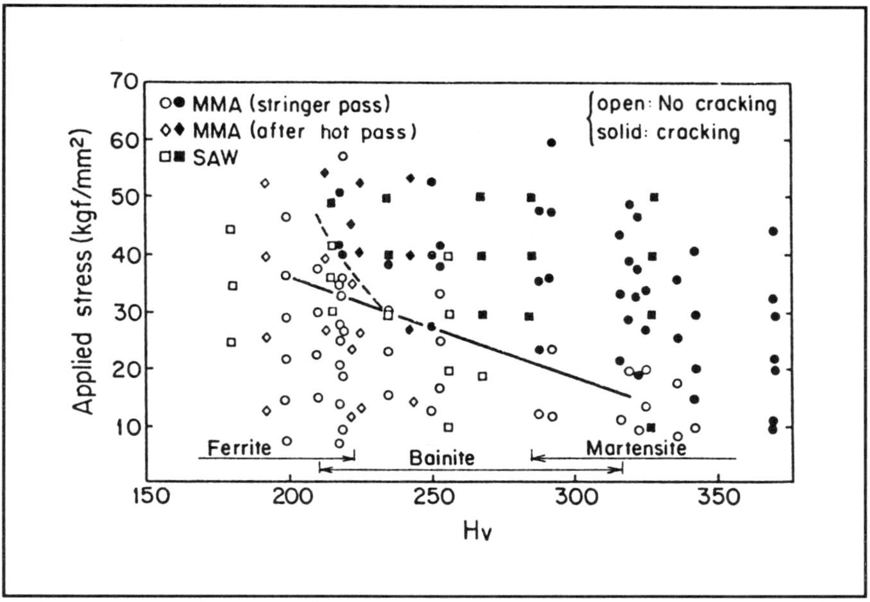

Figure 6–10 Relationship between the applied stress, microstructure and the maximum hardness for SSC of welds tested in the NACE solution (courtesy Sumitomo Metal Industries).

SSC, and they do, soft welds have also been observed to crack in sour environments. Thus SSC of welds is not solely a function of hardness. Fracture from SSC has been found in the base metal (BM), heat-affected zone (HAZ) and weld metal (WM) and not always associated with the hardest areas of the weldment. Above about HV260 there is a general correlation between SSC and hardness but below this fracture is less predictable.

Considerable work is being done in this area to better understand the SSC of welds in line pipe steels. However, at this time it is still considered good practice to stress relieve welds to reduce areas of high hardness and high residual stress, Figure 6–12. The cost of post-weld heat treatment (PWHT) on flowlines and pipelines can be prohibitive, so many companies use ASTM A 106 Grade B or API Spec 5L Grade B pipe with E6010 electrodes. The resulting hardness in the as-welded condition are sufficiently low that SSC has generally not been a problem. For high pressure flowlines, heavy wall Grade B pipe such as 1-in.-thick is commonly used, often with a preheat that helps reduce cracking.

One form of corrosion attack unique to ERW pipe is grooving corrosion. Grooving corrosion is the selective corrosion of the narrow weld line in ERW pipe that occurs as a result of the galvanic difference between the weld and the pipe

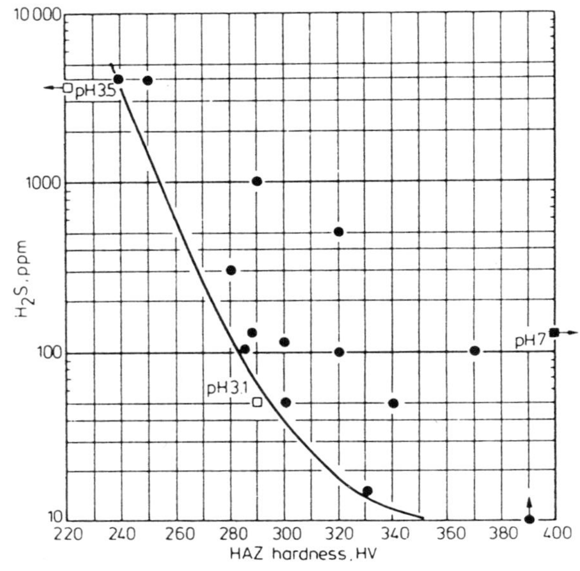

Figure 6–11 Maximum hardness of HAZ to avoid cracking in H_2S for pipline welds (courtesy The Welding Institute).

body. Figure 6–13 shows the cross section of a corrosion groove on the inside of a section of line pipe. Once the corrosion has progressed to the stage where sufficient wall thickness does not remain to contain the pressure, rupture occurs. This type of failure is frequently misinterpreted as the result of poor weld quality during manufacturing. Grooving corrosion is not dependent on grade or size and the susceptibility cannot be heat treated out or alloyed to correct the problem. Usually, lower carbon and sulfur content help reduce the susceptibility. Grooving corrosion almost always occurs in oxygenated high water cut systems, therefore, it predominates in surface piping, flowline, gathering lines, and water injection systems. Although grooving corrosion is most often observed internally, it has also been found on the external surfaces of line pipe. Since grooving is essentially a galvanic corrosion problem, the weld displays long narrow slits of corrosion along the weld while the remaining body of the pipe is often free of corrosion, Figure 6–14.

Another consideration in surface piping, flowlines, and gathering systems is corrosion from CO_2. Corrosion by wet CO_2 was discussed at length in Chapter

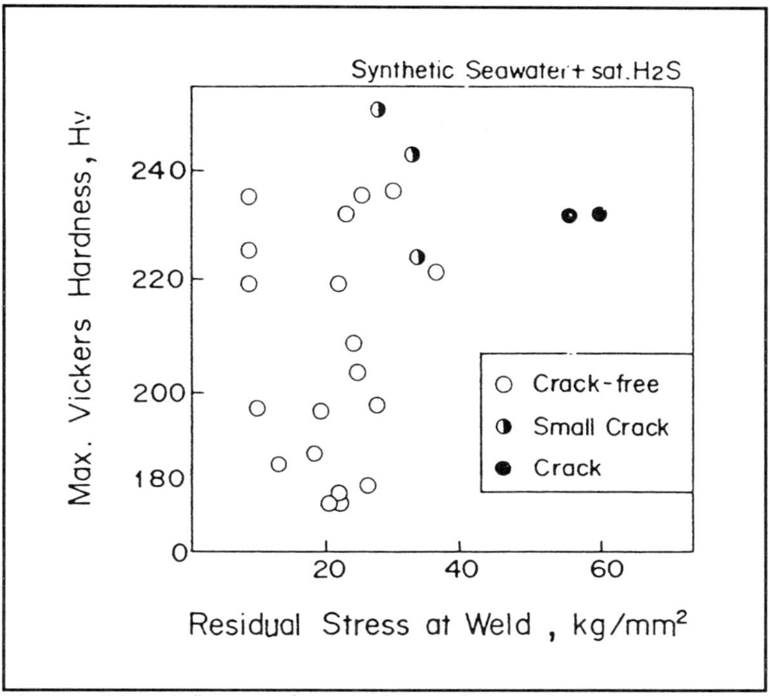

Figure 6–12 Influence of residual stress from welding and hardness on SSC of welds in line pipe (courtesy Sumitomo Metal Industries).

Four, however, particular importance must be paid to surface piping because of the selective corrosion of welds from CO_2. Even though all surfaces of steel components are susceptible to corrosion in the presence of CO_2, the welds are generally attacked at a higher rate due to the turbulence created by the weld bead itself. This turbulence reduces the ability to form a protective or semi-protective iron carbonate film on the weld that often forms on the remainder of the pipe surface. Therefore, the corrosion is accelerated at welds both because of the absence of a corrosion film and the turbulence created that also increases corrosion by erosion-corrosion or flow enhanced corrosion.

Of course, erosion-corrosion is not restricted to CO_2 systems; rather it can be a significant factor in any flowing system where corrosion is active. Erosion-corrosion is typically a greater problem in piping and line pipe compared to vessels because of the higher velocities.

The API has recognized that a critical velocity exists above which erosion-corrosion can become a significant problem. API RP14E provides an equation to

Figure 6–13 Grooving corrosion of the ERW in line pipe. The top of the photograph is the inside surface of the pipe.

calculate critical velocity (Vc) for design purposes so these velocities can be avoided. The equation:

$$V_c = \frac{C}{\rho^{1/2}}$$

where	V_c	=	velocity in ft/sec
	ρ	=	fluid density in lb/cu.ft.
	C	=	constant, 100–125

is strictly empirical and has been the source of much controversy. Recently an equation of similar form has been developed based by Craig on two phase flow that better defines the constant C. In the equation:

$$V_c = \frac{C}{\rho^{3/7}}$$

the constant C is dependent on corrosion product composition such that for carbon and low-alloy steels C = 90 when the corrosion film is composed of $FeCO_3$, C = 95 when it is FeS and C = 150 when the corrosion product is Fe_3O_4. In the case of chromium containing steels (≥ 13% Cr) the constant varies from 160 to 300 depending on the particular oxide that forms.

At velocities below V_c corrosion from CO_2 can still be a problem; however,

Figure 6–14 View of inside of the pipe shown in Figure 6–13. Note the distinct weldline attack in the center of the photograph (arrow).

erosion is not typically involved. At velocities in excess of V_c chemical inhibition is often impossible because the inhibitor cannot remain on the pipe surface without being stripped off by the flowing fluid.

Poor loading and shipping practices have led to numerous failures of pipe in transit. The cause of failure is fatigue and thus is referred to as transit fatigue.

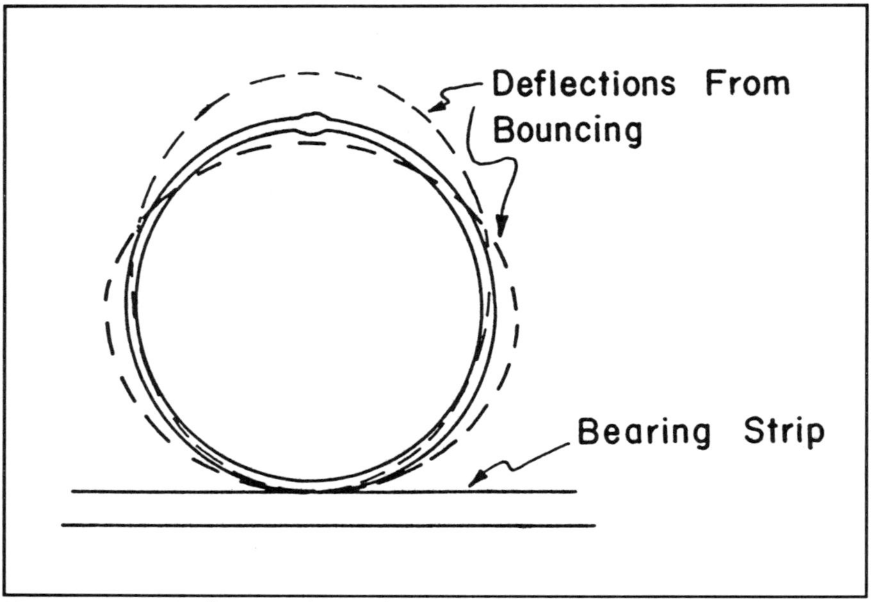

Figure 6–15 Deflection of pipe during transportation that results in fatigue loading (courtesy *Pipe Line Industry* and Gulf Publishing).

Transit fatigue results from gravitational and inertial forces that created cyclic stresses during transportation. An illustration of how these cyclic stresses arise is shown in Figure 6–15. The weight of the pipe provides a gravitational stress that fluctuates by bouncing and flexing as the pipe load moves up and down. As can be seen, the loads must be relatively high and there must be a large number of cycles to induce cracks, thus transit fatigue is not often observed when pipe is trucked. Most transit fatigue failures are associated with ship or rail transport where the pipe loads are higher and the trip is longer (higher number of cycles). Some of the other variables that influence transit fatigue are size of the contact area, nature of the bearing surface, amount of surface damage, the ambient environment, and the diameter to thickness ratio. The latter variable is expected because of the greater flexing with greater D/t ratios.

Surface damage such as denting, abrasion, fretting, and corrosion pitting can lower the fatigue resistance.

Transit fatigue failures originating in the pipe body are almost always associated with surface damage, whereas fatigue of submerged-arc welded pipe frequently initiates at the weld toe because of the change in cross section from the weld to the pipe body, Figure 6–16.

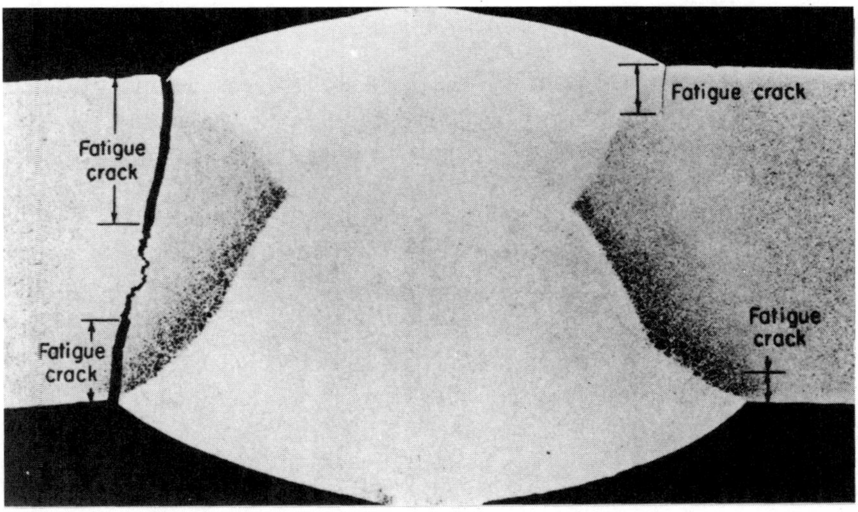

Figure 6–16 Fatigue cracks that initiated at the toe of the weld as a result of transit fatigue (courtesy *Pipe Line Industry* and Gulf Publishing).

Transit fatigue failures can be prevented by following good loading and shipping practices, for example those detailed in API RP5L1 and API RP5L5.

CRA Line pipe

Recently, two new methods of corrosion control using CRAs have been introduced: alloy clad line pipe and solid CRA line pipe. Clad line pipe is manufactured with carbon steel base metal and various CRA clad material such as 316 stainless steel, duplex stainless steel, and Incoloy 825. There are several different methods of manufacturing clad pipe that fall into two basic categories: lined and metallurgically bonded. The most common commercial methods for cladding pipe are: coextrusion, roll bonding, centrifugal casting, hydraulic shrink fit, and explosive bonding. The first three processes produce a metallurgical bond of the CRA material with the steel outer pipe. The latter two methods produce a lined product that results in an interference or mechanical fit but not a metallurgical bond at the interface.

In coextrusion, a piece of steel is initially extruded, then a block of CRA is introduced in this hollow and the two are extruded together so the CRA bonds to inside of the steel as the seamless tube is being made. Roll bonding is typically for larger diameters and is accomplished by rolling a thin sheet (~3mm) of CRA on a steel plate at elevated temperature. The plate is then formed similar to that described in the DSAW process and the pipe is long seam welded. In centrifugal casting, the outer steel shell is rotated while the CRA powder is introduced inside and cast against the ID. The hydraulic shrink fit is performed several different ways but the most common is to introduce a CRA pipe into a steel pipe that has already been heated to a moderate temperature (~800°F). The CRA pipe is hydraulically expanded against the steel pipe as it cools, thereby providing a shrink fit. Another method for lining pipe is to insert a CRA liner into a steel pipe and then set off an explosive charge inside the CRA liner to expand it against the steel. Both methods of interference lining create a residual stress on the order of 15,000 psi or greater, thus preventing the liner from easily slipping out of the outer tube. Clad pipe has not been extensively applied to date; however, the increased welding rates of clad lines is approaching carbon steel pipeline welding and thus clad lines are expected to be a frequent and cost effective alternative for the future.

Solid alloy pipelines and flowlines have been gaining wider acceptance and as more lines are constructed, welding speeds are increasing. However, the lines laid to date have been duplex stainless steel for high CO_2 service and solid Hastelloy G-3 for gas containing appreciable H_2S. It is expected that as welding techniques and speeds improve, alloys such as Incoloy 825 will be used for H_2S service. Welding consumable selection is also very important to the welding and performance of alloy line pipe.

Of course, there are many factors that must be considered in the selection of CRA line pipe. Besides CO_2, H_2S, and chloride content of the fluids, surface piping typically contains oxygen so the environment is often more severe than downhole. Moreover, welds present another concern from a corrosion standpoint. Materials such as duplex stainless steels require special care to ensure the balance of ferrite and austenite is sufficient for maximum corrosion resistance.

Field Separators and Storage Tanks

Separators and tanks are designed and built to various codes and standards depending on their intended use. For pressure containing vessels such as separators the ASME Pressure Vessel Code is used for design requirements. For storage tanks, API Standards 620 and 650 apply. Pressure vessels such as two- and three-phase separators are typically manufactured from ASTM (ASME) A 106, A 285, A 516, Grade 60 or 70, A 517, A 537, etc. Likewise storage tanks are also constructed from these steels as well as ASTM A 36.

As with pipelines, several forms of hydrogen damage can occur in vessels and tanks. Original fabrication welds on the vessel for construction of the vessel or attachment of nozzles can produce hard welds that are susceptible to SSC. In addition, high residual stresses from welding can lead to SSC even in soft welds. The petroleum refining industry recognized this latter problem in the early 1970s.

The problem was most frequent when carbon steel vessels were submerged-arc welded (SAW) using bonded (alloying) fluxes and put in wet H_2S service in the as-welded condition. Transverse-weld cracking was found to occur at hardness levels above and below the previously established HB 236 hardness (HRC 22) criteria.

API Recommended Practice 942 and *NACE Recommended Practice RP-04-72* cover welds for carbon steel refinery equipment and contain discussions of the various factors affecting weld hardness and agree on a mandatory maximum weld hardness of HB 200. The API RP 942 places additional mandatory restrictions on weld metal chemistry. The limits for carbon, manganese, and silicon are based upon the P-1 chemistry limits of ASME Code Section IX, Table Q11.3, while the total content of chromium, nickel and molybdenum is set at 0.25%. It should be pointed out, however, that these two documents are by definition intended to cover carbon steel (P-1) material only and do not address steel such as Type SA 612 steel, which contains higher manganese and silicon than the more common SA 516 Grade 70 steel, but also allows additional small amounts of nickel and chromium for strengthening and grain refinement.

Failures in carbon steel vessels are associated with high hardness and high manganese resulting from multipass welds using bonded fluxes. The additive effect of the alloying fluxes results in manganese contents often exceeding 2% and weld strengths, as measured by hardness, which are disproportionately high for unalloyed steel. When using bonded fluxes, it is not unusual to find weld deposits having over HB 225 hardness in a carbon steel Type SA 516-70 pressure vessel.

All vessels and tanks intended for sour service should be post-weld heat-treated to remove residual stresses from welding and reduce the hardness to BHN 200 or less.

Another serious problem in corrosive service is the fabrication of dissimilar metal welds, especially between carbon or low-alloy steels and austenitic stainless steels. Catastrophic failures have occurred in these welds by SSC and SCC. Figure 6–17 shows the cracking of a dissimilar metal weld between low alloy steel plate and the austenitic stainless steel filler metal from SSC. In these type of welds, chromium diffuses from the austenitic filler metal to the fusion line and carbon diffuses to the same region from the low-alloy steel side. The two

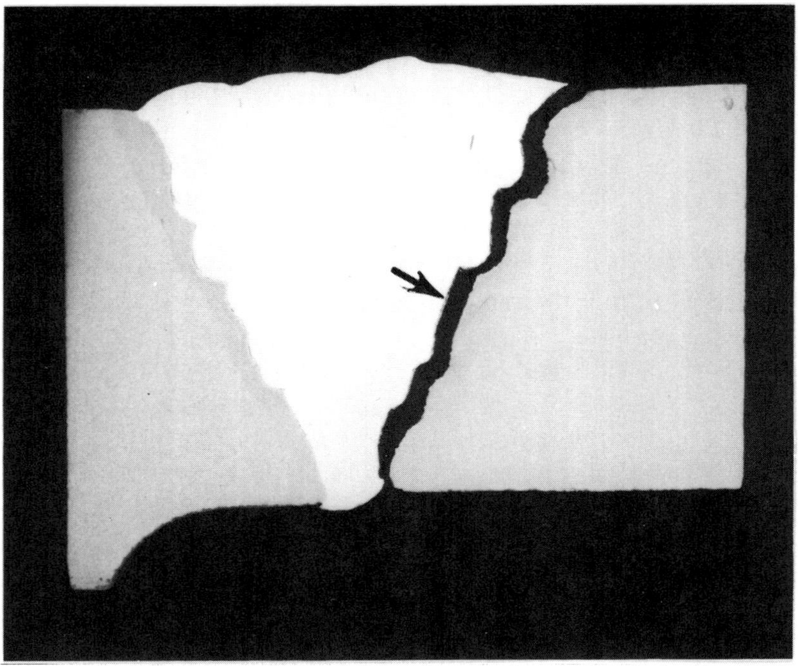

Figure 6–17 Cross section of dissimilar metal weld that failed by SSC.

elements (and other carbide formers if they are present, i.e., Ti, W, Mo, etc.) combine to form a hard, narrow, chromium-carbide-rich martensite phase approximately 0.002 in. wide along the fusion line, Figure 6–18. The hardness in this zone can exceed HRC 40. This hard zone cannot be post-weld heat-treated (PWHT); in fact PWHT makes the situation worse by promoting additional diffusion of Cr and carbon. Therefore, dissimilar metal welds are discouraged in very corrosive environments, especially if H_2S is present, because of the potential for catastrophic cracking failures.

Other forms of hydrogen damage such as blistering and HIC are often experienced by vessels in sour service. As discussed before, these forms of damage are not related to hardness but to the number and shape of inclusions, the H_2S content and pH of the water. HIC of production facility vessels has become such an industry concern that many companies now require NACE TM-02-84 tests on plate material before construction of vessels. These tests are performed in either the standard TM- 02-84 solution or modified by testing in the TM-01-77 solution.

One system of particular interest is glycol dehydration, specifically within the contactor. Dehydration of gas is important from both a corrosion and operation

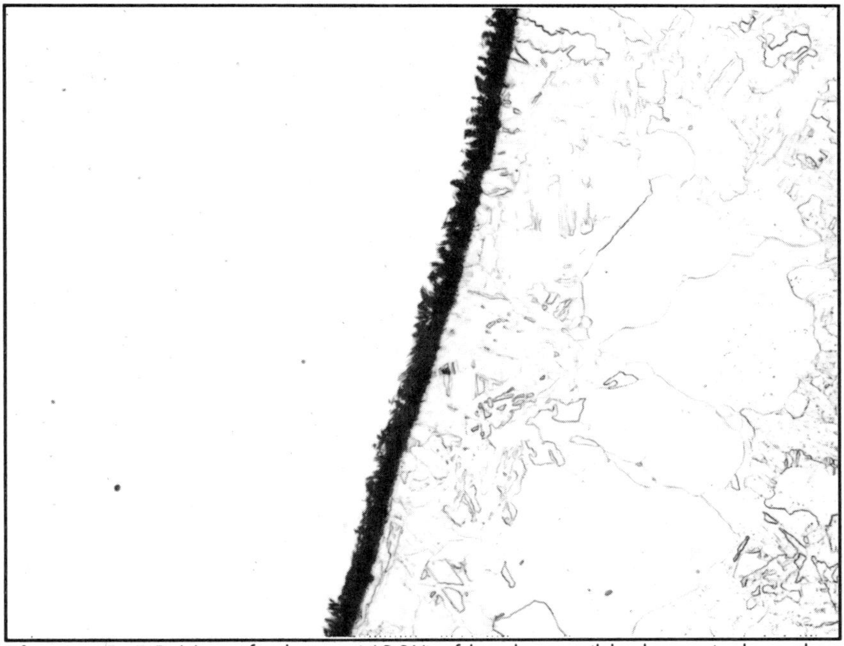

Figure 6–18 Magnified view (400X) of hard zone (black area) along the fusion line of the weld shown in Figure 6–17.

standpoint. If water or moisture is removed from the gas stream to below the dew point at the lowest expected temperature of downstream operations, then corrosion should be eliminated. However, it has been found in practice that unless the gas is dehydrated to 3–4 pounds of water per million cubic feet it will still condense water and corrosion will occur.

Wet gas enters the inlet scrubber where a pressure drop allows some water and condensate to drop out of the gas. The gas phase then enters the bottom of the absorber tower, which is filled with trays. As the gas rises in the absorber, it is contacted with lean glycol, which absorbs the water from the gas, becoming rich glycol. The dry gas then exits the top of the tower. The rich glycol comes off the bottom of the absorber tower and proceeds through a heat exchanger, then into the stripper (reboiler or regenerator). In the stripper, the glycol is heated so water is driven off to the top of the stripper and into the condenser and accumulator. The glycol moves out of the bottom of the stripper/reboiler as lean glycol for recycle of the process. Corrosion frequently occurs in the contactor because of the wet gas and the presence of CO_2 and H_2S. Contactors are often coated with a plastic coating or clad with a CRA when serious corrosion is expected.

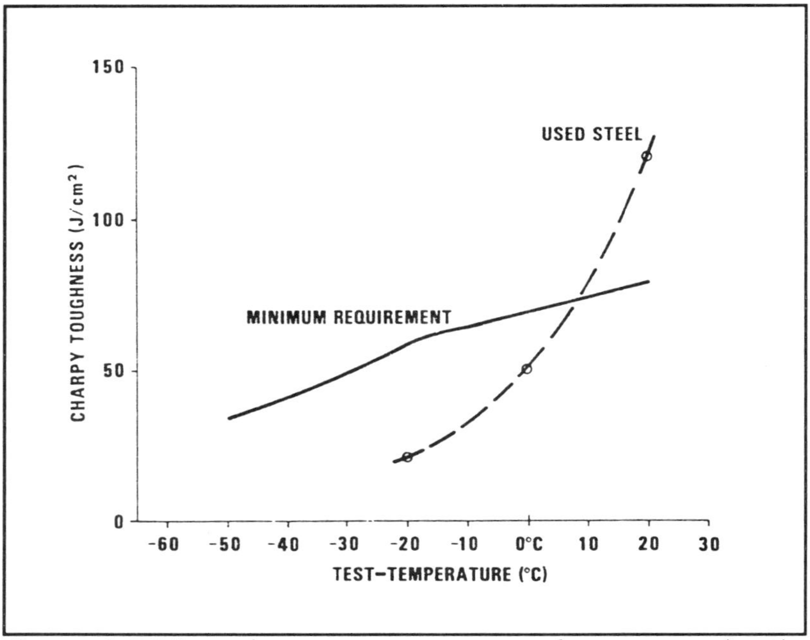

Figure 6–19 Degradation in impact properties of pressure vessel steel after sour gas service (courtesy G. Wilken and NACE).

During reboiling of a glycol, it must not be heated to its boiling point to remove water; otherwise, decomposition of the glycol may occur. These decomposition products are acidic organic compounds that are quite corrosive to steel. Even though the bulk temperature of the glycol may be below the decomposition limit, high metal skin temperatures can rapidly decompose glycol. To maintain low skin temperatures, proper control of heat flux by design and proper operation of reboilers is essential.

Degradation of glycols will lead to a lower pH of the solution and enhanced corrosion. The presence of oxygen can also lead to significant degradation of glycols again resulting in low pH solutions.

The addition of H_2S and CO_2 aggravates this corrosive condition and/or creates additional problems. Corrosion from these acid gases will be greatest whenever water drops out of the gas, such as in the inlet scrubber or water knockout vessel. Also, regeneration of the rich glycol creates an environment of high temperature water or vapor that contains concentrated H_2S and CO_2. Thus reboiler/stripper vessels may experience severe corrosion in the hotter sections and in the event of high H_2S blistering or HIC of the vessel steel.

Inlet scrubbers or free water knockouts have been found to contain blisters

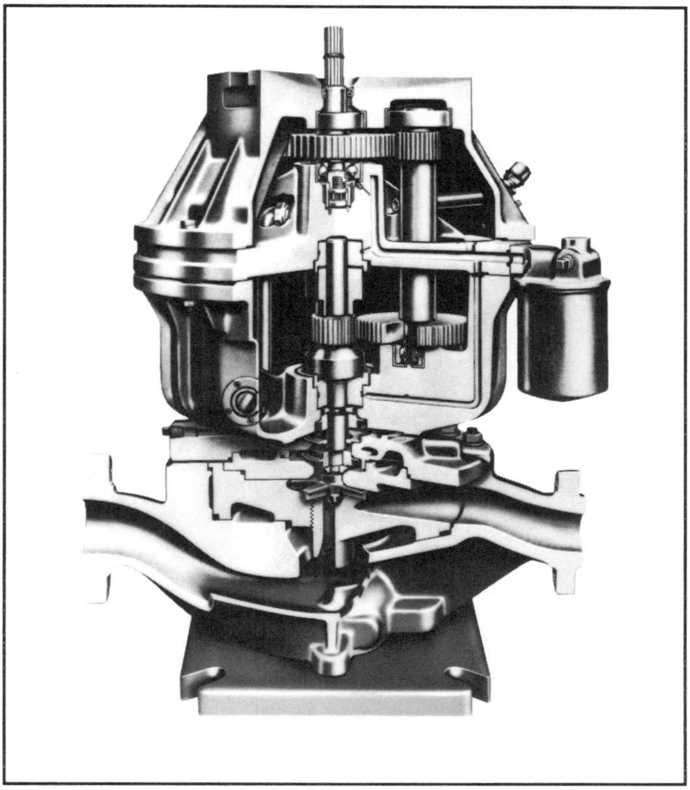

Figure 6–20 Cutaway view of a high-speed centrifugal pump (courtesy Sundstrand Fluid Handling).

after exposure to sour water. Blisters are frequently associated with nonmetallic inclusions (i.e., MnS, Al_2O_3) and predominate in banded steels. Blisters have been found to degrade the impact toughness of steels compared to their original properties, Figure 6–19.

Compressors and Pumps

There are numerous types of compressors and pumps used in gas and liquid transport and processing. Common compressors are: centrifugal, axial, positive displacement, and reciprocating. Pumps also come in the same general configuration.

Figure 6–20 shows a common design of a centrifugal pump. Fluid flows from

TABLE 6-4 Typical Centrifugal Pump Component Materials

	Service			
Component	**Crude Oil (Low Water)**[1]	**Water**[2]	**Crude, Condensate Mildly Corrosive**[3]	**Crude, Condensate Very Corrosive**[4]
Casing	CS	NiAlB, 410SS 316SS, DSS	CS, 316SS, DSS	316SS, Monel
Impeller	CS, CI	NiAlB, 410SS 316SS, DSS	316SS, DSS	316SS, Monel
Wear Rings	CI	HFSS	CI, HFSS	HFSS, Monel Nitronic
Shaft	CS, 4140	4140	4140	4140, 316SS
Shaft Sleeve	SS, HFSS	HFSS, Stell	HFSS, Stell	HFSS, Stell
"O" Rings	Teflon	Viton, Nitrile, Teflon	Teflon	Teflon

Footnotes 1 - For crude oil containing water less than about 30% by volume.
2 - Not containing significant amounts of H_2S.
3,4 - The user must arrive at a definition for mildly and very corrosive. Moreover, if sufficient H_2S is present in either case materials may also need to meet the requirements of NACE MR-01-75.

Materials - CS = Carbon Steel, (eg. 1045)
SS = Stainless Steel
DSS = Duplex Stainless Steel
HF = Hardfaced
Stell = Stellite
NiAlB = Nickel Aluminum Bronze
CI = Cast Iron

left to right through the path designated. Typical material choices for different components of a centrifugal pump are shown in Table 6–4. In systems that are mildly corrosive, carbon steel ductile iron or 316 stainless steel are satisfactory. However, as the severity of corrosion increases, those surfaces contacted by the process stream need to be more corrosion resistant, requiring the application of high alloy materials. These alloys often include duplex stainless steels, Monel 400, Hastelloy C, and Ti alloys. The need for this significant increase in alloying is not entirely due to corrosion but also to resist erosion damage, which may be quite severe in fluid and gas handling systems. Pumps are usually specified in accordance with the following API or ANSI Specifications:

API 610, Centrifugal Pumps for General Refinery Service

ANSI/ASME B73.1M—1984, Specification for Horizontal End Suction Centrifugal Pumps for Chemical Process

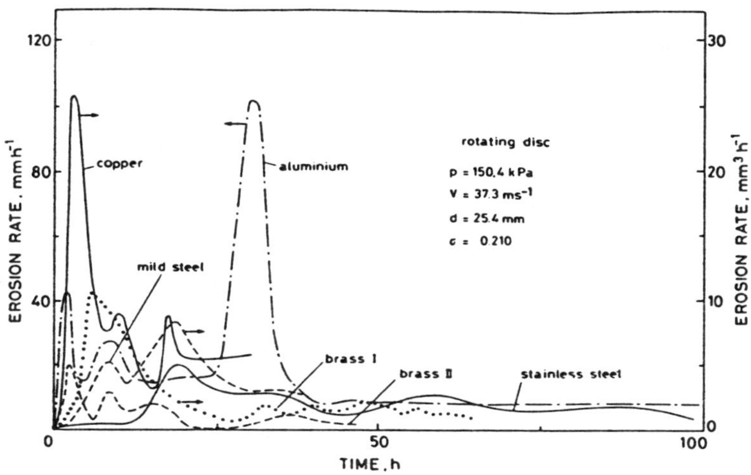

Figure 6–21 Erosion rate versus time for materials in a rotating disk test (courtesy ASM International).

ANSI/ASME B73.2M—1984, Specification for Vertical In-line Centrifugal Pumps for Chemical Process.

In the same manner as pumps, compressors are subjected to corrosive gas streams often at high velocity so erosion-corrosion can be quite severe. However, compressors handling dry gas can be made of less corrosion-resistant alloys if the gas temperature exceeds the dew point during operation and water is not allowed to collect during shutdown.

For relatively benign environments, impellers are often manufactured from carbon steel or cast iron. However, where high strength is needed, 4320 or 4330 low-alloy steels are used. In the presence of H_2S, these impeller materials must be reduced in hardness to resist SSC and meet NACE MR-01-75. This can be a problem since these components must have sufficient design strength to operate under high loads and resist erosion. As Figure 6–21 shows, erosion is best handled by the use of higher alloy materials such as stainless steel; however, it must be determined whether solid particles are also involved since they exacerbate erosion, Figure 6–22; and if cavitation may occur, Figure 6–23. As evidence by these figures, stainless steels, nickel-base alloys and cobalt base alloys provide the best erosion/cavitation resistance. Likewise, corrosion resis-

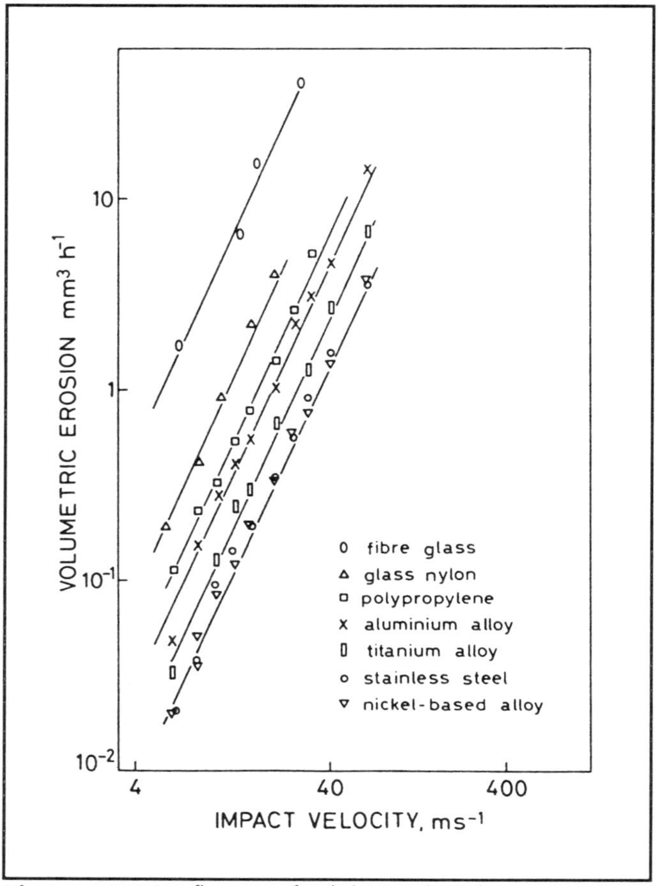

Figure 6–22 Influence of solid particle velocity on erosion rate of different materials (courtesy ASM International).

tance and resistance to both SCC and SSC are achieved through alloying. Intermediate alloys such as Ni-resist and 410 stainless steel are often successfully used in pumps and compressors in relatively mild service. Under no conditions are copper alloys (brasses, bronzes, etc.) acceptable for service in H_2S either for gas or liquid service. For the most severe service (hot H_2S, CO_2), Ti impellers, inducers, and diffusers have been found to perform excellently. In water injection and WAG service copper base alloys have been quite successful.

One other failure mode often observed on compressor impellers is fatigue and/or corrosion fatigue. The high alternating stresses imposed on the blades of the impeller provide easy development of fatigue cracks. As illustrated in Chap-

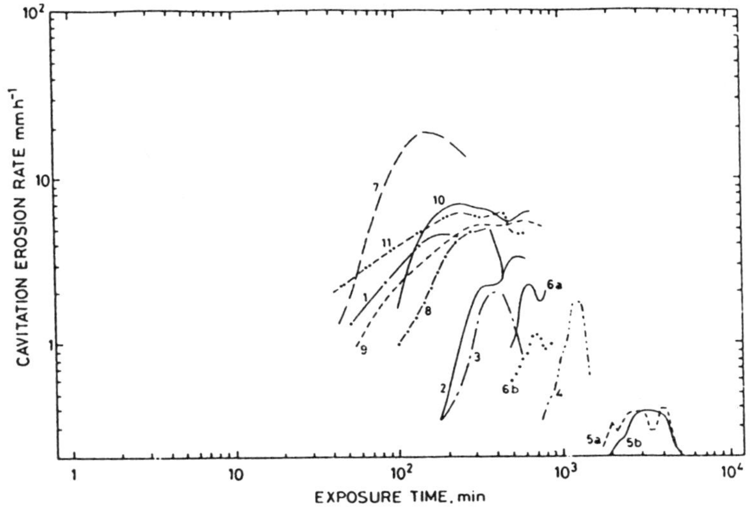

Figure 6–23 Cavitation-erosion rate of alloys versus time (courtesy ASM International). 1 X15–5 CrVMo121; 2 Hastelloy B–2; 3 Hastelloy C–4; 4 Haynes alloy 25; 5Haynes alloy 6B; 6 Mn steel; 7 41 Cr Mo S4; 8 Gx–CrNiMo 18–10; 9 X5 CrNiCuNb 17–4; 10 Gx–CrNiMoCu 25–6; 11 G–CuAl9NiFe.

ter Two, the introduction of a corrosive environment will result in the loss of a fatigue threshold so that failure eventually results regardless of stress level.

Valves

A variety of valves are used in surface facilities to regulate flow of gas, liquids, and combined fluids. Some of the more common valves are gate, ball, globe, plug, butterfly, needle, and check. The above names describe the type of internal feature that controls flow.

All valves have certain similarities and differences. However, in all cases the valve body must act as a pressure vessel and as such must be resistant to fracture under service conditions and leakage through the body where it has been intruded by the stem, fittings, etc. Valve bodies may be forged, cast, or machined from wrought material. The majority of valves are carbon or low-alloy

TABLE 6–5 Typical Materials for Various Valves

		Service			
Valve Type	**Component**	**Crude Oil (Low Water)[1]**	**Water[2]**	**Mildly Corrosive[3] Crude, Condensate, Gas**	**Very Corrosive[4] Crude Condensate Gas**
Gate	Body & Bonnet	CS	CS, 13Cr	CS, 316SS	316SS, Monel,
	Stem	CS	13Cr, 316SS Monel	316SS, Monel	Inconel 718 Inconel 925
	Gate	CS/13Cr	CS, 13Cr, 316SS, Stell	13Cr, HFSS 17-4PH	HFSS, HF Inconel 718
	Seat	CS/13Cr	CS, 13Cr, Stell	13Cr, Stell	Stell
	Packing	Teflon	Teflon Graphite Filled	Teflon	Aflas, Kalrez
Ball	Body & Bonnet	CS	CS, 316SS, Monel	CS, 316SS	316SS, Monel Inconel 625
	Stem/Trunnion	CSP, 316SS	316SS, Monel	CS, 316SS	Monel Inconel 718
	Ball	CSP8 316SS	CSP, 316SS, Monel	CSP, 316SS	Monel, 316SS, Inconel 625
	Seat Holder	CSP	CSP, 316SS	316SS, Monel	Monel
	Seat Insert	Teflon, Viton	Teflon, Viton	Peek, Teflon, Viton	Peek, TFE
	Seals	Teflon, Viton	Teflon, Viton	Teflon Viton	Aflas, Kalrez
Plug	Body & Bonnet	CS	CS, 316SS, DSS	CS, 316SS, DSS	(5)
	Stem	CS	316SS, 17-4PH	CSP, 316SS, DSS	
	Plug	CS	316SS, 17-4PH	316SS, 17-4PH	
	Seat	CS	CSP, 316SS	316SS, CSP	
	Packing	Graphite Filled	Grapite Filled	Graphite Filled	
Butterfly	Body	CS	CS, 316SS	CS, 316SS	(6)

TABLE 6–5 (continued)

Valve Type	Component	Service: Crude Oil (Low Water)[1]	Water[2]	Mildly Corrosive[3] Crude, Condensate, Gas	Very Corrosive[4] Crude Condensate Gas
	Stem	CSP, 17-4PH	17-4PH, Monel	17-4PH, Monel	
	Disc	CSP, 316SS	316SS, Monel	316SS, Monel	
	Seat	316SS	Reinforced Teflon, 316SS	Reinforced Teflon, 316SS	
	Packing	Teflon	Teflon Graphite Filled	Teflon, Graphite Filled	
Check	Body & Bonnet	CS	CS, Brass	CS	316SS, Monel Inconel 625
	Clapper	CS, 316SS	Brass, 316SS	316SS	316SS, Monel Inconel 625
	Face Seal	Viton, Buna	Viton, Buna	Viton, Buna	Teflon
	Pin	13Cr, 316SS	316SS	316SS	Inconel 718
	Seat	CS, 316SS	316SS	316SS	Monel
	Seals	Viton, Buna	Viton, Buna	Teflon	Teflon

Footnotes 1 - Crude oil with water cuts less than 30% by volume
2 - Not containing significant amounts of H_2S.
3,4 - The user must arrive at a definition for mildly and very corrosive. Moreover, if sufficient H_2S is present in either case, materials may also need to meet the requirements of NACE MR-01-75.
5,6 - These valves are not typically used in very corrsive fluids, therefore resistant alloys are not available.

Materials: CS = Carbon stee
CSP = Carbon steel plated with hard chrome or electroless nickel
SS = Stainless steel

steel. The low-alloy steels are most frequently used when increased fracture toughness at low temperature is required. Increasing corrosion resistance is achieved by changing to stainless steels, bronzes, nickel base alloys, or titanium alloys. Also, CRA clad valve bodies are available for corrosive service.

Gate valves are described by *API Specification 6A* and pipeline valves by *API Specification 6D.*

Besides the bodies, many valves have stems, springs, and seats in common. These internals, along with the gate, ball, plug, etc., are also manufactured from a variety of materials, depending on the service. Typical materials for the above components are presented in Table 6–5. While these do not represent all of the possible materials available they do illustrate some of the more common materials.

Care must be taken in the selection of materials for valves to ensure all components of the valve are suitable for the intended service.

Heat Exchangers

The three most common heat exchangers are shell and tube, air coolers, and plate. Shell and tube will be presented as representative of the other types although there are certain peculiarities for each type of exchanger. Heat exchangers are frequently constructed of carbon steel for most corrosion free or moderately corrosive environments. However, as the corrosivity of fluids on the shell side and/or tube side increases the need for alloying becomes greater. Besides the corrosivity of the fluid, design of the exchanger is critical to a long service life since hydrodynamics play an important role in corrosion and erosion. Thus the shell side will often have low flow and dead spot areas where corrosion may be typified more by static conditions rather than the dynamic conditions which are characteristic of tube sides.

As the corrosivity of the fluid increases, steps such as plastic coating or, for more severe environments, CRA cladding of exchanger shells may be required. The selection of specific alloys for the shell or cladding is dependent on the particular process stream, temperature, pressure etc. Shell material depends on the specific service conditions and may range from carbon steel, to clad CRAs on steel, to solid corrosion-resistant alloy shells. The decision between cladding and a shell constructed of solid CRA is determined primarily by the exchanger shell dimensions. Small vessels are more economically constructed of solid CRA compared to cladding or weld overlaying. Larger diameter vessels on the other hand are more suited to cladding and much more economical than solid alloy vessels.

In the event streams on the shell side will contain H_2S, the shell material and welds should be manufactured and specified in accordance with NACE MR-01-75. All welds should be stress relieved to reduce both hardness and residual welding stress. Likewise higher alloy material should be made in compliance

TABLE 6–6 Pitting Resistance Equivalent of Some Common Stainless Steels

ALLOY	PRE
Type 405	13
Type 304	18
Type 316	24
2205 Duplex	34
Alloy 28	39
AL6X	42
254 SMO	43
2507 Duplex	43

with MR-01-75 when elevated concentrations of H_2S are expected. The same is true for the tube side but often the erosion-corrosion aspect is also important in selecting tube material. Tubes carrying seawater or chloride-containing water can often be handled successfully with stainless steels as long as the flow rate is high enough to avoid prolonged contact of the fluid with the tube wall. Moreover, it is necessary for exchanger tubes to be thin walled to permit maximum heat exchange, which limits the use of a corrosion allowance and requires corrosion resistant alloys. Duplex stainless steels have been successfully used for exchanger tubes where H_2S levels are relatively low. Since pitting resistance of stainless steel and nickel alloys from chlorides is a function of temperature, many alloys will be limited by temperature, Figure 6–24. This figure illustrates the benefit of increasing molybdenum content on pitting resistance.

Pitting resistance of stainless steels can also be related to their chemical composition by the following regression equation:

$$PRE = (\text{wt \% Cr}) + (3.3 \text{ wt\% Mo}) + (16 \text{ wt} + \text{\% N})$$

A comparison of the Pitting Resistance Equivalent (PRE) for different alloys is shown in Table 6–6. The strong effect of nitrogen in the above equation can be seen by comparing Alloy 2507 that contains 0.3%N to Alloy 2205 that contains essentially no N. The real value of this equation is for comparison of the multitude of stainless steels of different chemical composition where no corrosion data are available. It does not predict the actual corrosion rate.

Increasing temperature, velocities, the presence of dissolved oxygen and H_2S can overwhelm stainless steel and Incoloy tubes (e.g., 825, 800, etc.) so that titanium tubes are sometimes used for very aggressive streams at higher temperatures. Ti heat exchangers have been successfully used in the refinery and chemical process industry for many years and have been used for shell and tube seawater exchangers in the North Sea.

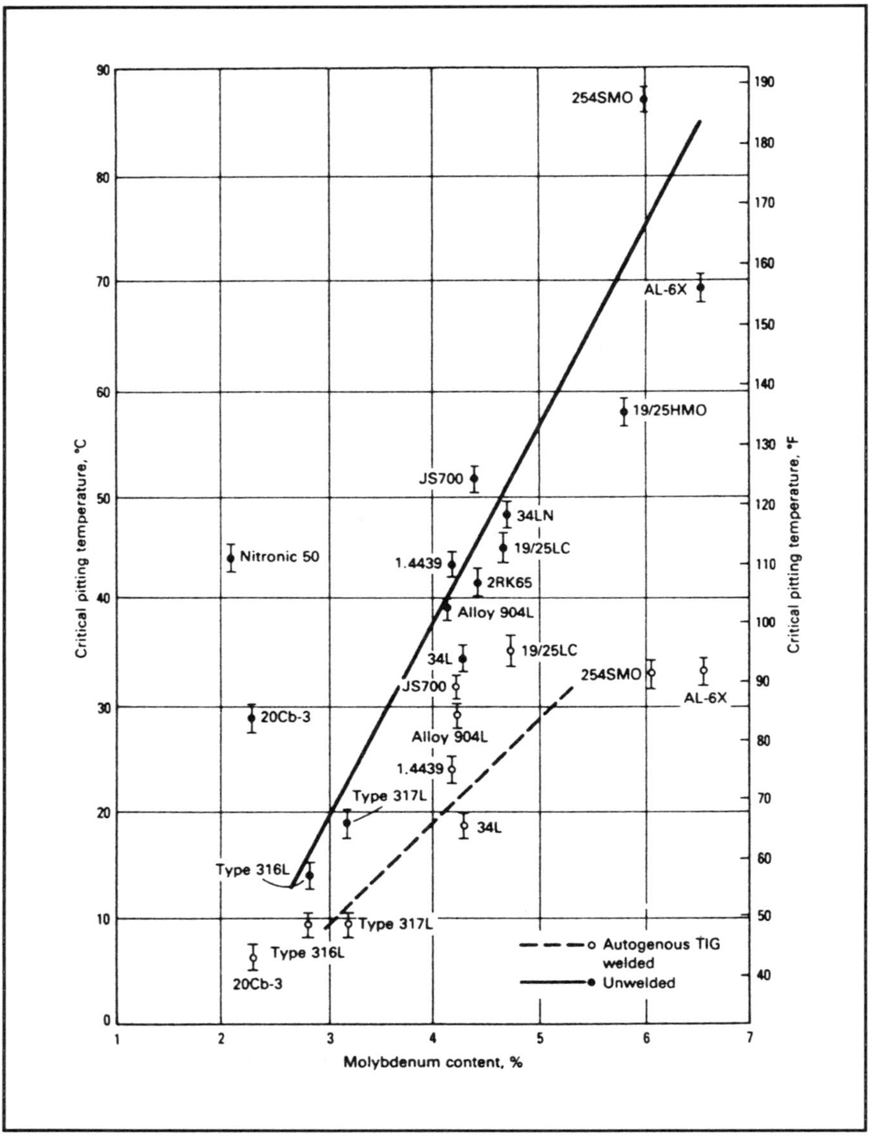

Figure 6–24 Critical pitting temperature versus molybdenum content of various welded and unwelded austenitic stainless steels (courtesy ASM International).

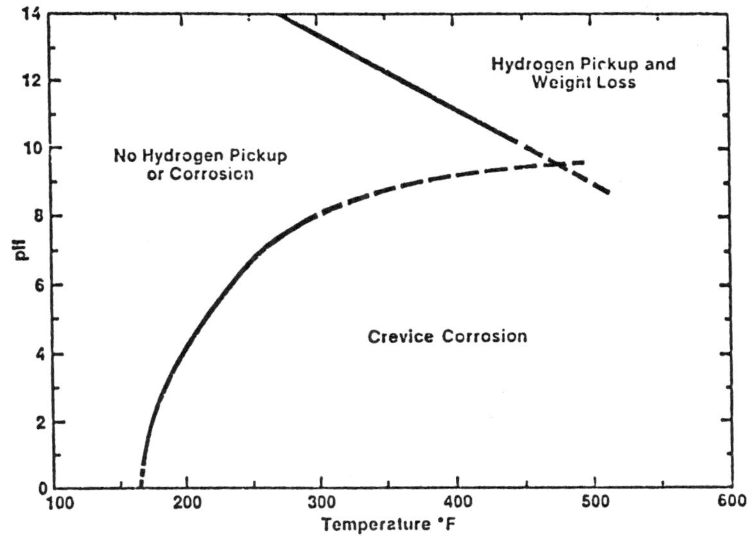

Figure 6–25 Effect of temperature and pH on crevice corrosion of unalloyed titanium (ASTM Grade 2) in saturated brine (by permission of AIChE,© 1978;After Covington, L.C, R.W. Schutz and I.A. Franson, AIChE Meeting, New York, November 1977).

As with any alloy, there are some drawbacks to titanium. Titanium does have temperature limitations, becoming susceptible to crevice attack with increasing temperature, Figure 6–25. Hydriding of tubes and tube sheets has been reported when titanium is galvanically connected to less noble metals such as steel.

The selection of materials for heat exchangers is similar to that for all the other surface equipment and requires a specific knowledge of the process stream, temperature, pressure, etc., to make the correct choices.

References

API Specification 5L. "Specification for Line Pipe." 1991.

Bruno, T. V. "How to Prevent Transit Fatigue of Tubular Goods." *Pipeline Industry*, (July 1988), p. 31.

Couper, A. S., and A. Dravnieks. *Corrosion*, 18 (1962), p. 291t .

Garner, A. "How Stainless Steel Welds Corrode," *Metals Progress*, 27, (April 1985) p. 32.

Gas Engineers Handbook, 51 (1968) p. 8.

Glenny, R. J. E., J. E. Northwood and A. Burwood-Smith. *International Metals Review 193*, 20 (1975), p. 1 .

Gooch, T. G. *The Welding Institute Research Bulletin*, (Aug. 1982), p. 241.

Gray, M. and A. B. Rothwell, "How Welding Affects Pieline Steels." AWS Conference, Houston, TX (Feb. 1980).

Ikeda, A., T. Kaneko and F. Terasaki. Paper No. Eight. Corrosion/80, National Association of Corrosion Engineers (1980).

Ikeda, A., et al. *Sumitomo Search*, No. 26 (1981), p. 91.

Kaneko, T., M. Takeyama, M. Nakanishi, Y. Sumitomo, and A. Ikeda. "Improvement of Hydrogen Sulfide Cracking Susceptibility in Line Pipes for Sour Gas Services." Middle East Corrosion Conference, National Association of Corrosion Engineers (1979).

Karimi, A., and J. L. Martin. *International Metals Review*, 31 (1986), p. 1.

Nakai, Y., H. Kurahashi, N. Totsuka and Y. Wesugi. "Effect of Corrosive Environment on Hydrogen-Induced Cracking." Corrosion/82, National Association of Corrosion Engineers, Houston, TX (1982).

Sato, E., M. Hashimoto, R. Matsuhashi, and T. Murata. "Corrosion of Steels in a Wet H_2S and CO_2 Environment," Second Asian-Pacific Corrosion Control Conference, Kuala Lumpur, (1981).

Sundstrand Fluid Handling Brochure,(1984).

Tanaka, T., Y. Ito, M. Nakanishi, T. Kaneko and Y. Komizo, *Sumitomo Search*, 23 (May 1980) p. 27.

Wilken, G. Paper No. 158. Corrosion/83, National Association of Corrosion Engineers (1983).

Yamada, K., et al. Paper No. 8306-032. 1983 ASM Metal Congress, Philadelphia, PA (1983).

7

FAILURE ANALYSIS

This chapter deals with failure analysis. The analysis of failures is important for several reasons. First and foremost, understanding the cause of a failure of a component can often lead to correction of the problem and prevention of a recurrence of the failure. Other advantages of a failure analysis are: a better understanding of the actual operation of a piece of equipment versus its intended use or design operation; recognition of upset conditions in the system that were not anticipated; evaluation of the performance of maintenance programs; and corrosion control program and legal liability.

The important stages of a failure investigation will be presented and discussed; however, failure analysis is a complex field, quite often involving an interdisciplinary approach that may include metallurgy, analytical chemistry, surface science, stress analysis, fracture mechanics and a host of other disciplines. Therefore, the discussion that follows is but a brief introduction to the subject.

The primary stages of a failure analysis are as follows: however, the sequence may vary depending on the particular investigation and may not involve every step.

1. Retrieval and security of as many pieces of the failed component as possible.
2. Preliminary visual examination and identification of all pieces of the failed component.

3. Cataloging of all the components.
4. Background data collection.
5. Planning of tests to be performed and samples to be selected.
6. Removal of samples for chemical analysis (e.g., corrosion products, deposits, coatings, etc.) that may be affected by further work.
7. Nondestructive evaluation (NDE).
8. Macroscopic examination.
9. Removal of samples for destructive testing and preservation of all remaining pieces.
10. Mechanical testing.
11. Microscopic examination, including metallography, electron microscopy.
12. Chemical analysis of bulk components (alloys and surface analysis of components (scale, corrosion films, etc.).
13. Stress and fracture mechanics analysis.
14. Review of all the data and formulation of a scenario for cause(s) of failure.
15. Re-confirmation that the scenario and mechanisms for failure are consistent with the data.
16. Report and recommendations to prevent the failure in the future.

Retrieval of Failed Components

The retrieval of all or as many of the failed pieces of equipment is of the utmost importance in ascertaining the cause of a failure. Every effort should be made to obtain the original failure and not just similar components for analysis. In the event a sample is to be forwarded to a laboratory for analysis, the engineer performing the analysis should be contacted to inquire as to the desired pieces for analysis and how they should be shipped and handled.

Preliminary Visual Examination and Cataloging Components

Before performing any work, a preliminary visual examination of all the pieces of the failure should be carried out. Moreover, all pieces should be identified for future reference. Frequently, this stage is best suited to macrophotography to provide a photographic record of the condition of the failure and the pieces received for examination. At this time dimensional analysis is also performed to obtain data that may be necessary later on in the investigation for reconstruction of the failure.

When photography is not acceptable, such as in a very large catastrophic failure, or the conditions do not allow for photography, then a schematic or sketch should be made with critical dimensions included.

In cases where numerous pieces of a failure exist, it is often necessary to produce a detailed list and description of all components of the failure to ensure they are accounted for at every stage of the investigation. Furthermore, in cases of litigation, handling by countless people and years of legal action may ensue before issues are resolved. This accountability of all components is a major responsibility during any failure analysis.

Background Data Collection

Collection of background data is a very important part of a failure evaluation. This includes information regarding the manufacturing of the components, the standards it was designed and built in accordance with, the way it was actually operated in service, the nature of the service environment, and any upsets in the normal operations. Many failures occur as a result of upset conditions that were not anticipated when the equipment was designed or selected. Any and all of this information is of value in determining the cause of a failure. Contrary to the thinking of some unfamiliar with failure analysis, the cause of a failure often cannot be solely determined from the failure itself, especially when competing mechanisms may be equally possible. Operational history can often sway the final determination of the cause simply because failure by another mechanism could not occur under those operational conditions.

Planning of Tests and Sample Selections

Before any destructive work is performed, planning the tests to be accomplished can save time and money as well as reduce the risk of damaging necessary evidence that may not be readily apparent. Time spent in planning and sample selection is time well spent in the investigation but most often is the stage that is carelessly undertaken or eliminated because of pressure to perform a quick analysis.

In the case of fractures, samples must be selected such that a minimum of damage occurs during sample removal so that if additional samples are necessary later in the investigation there are undamaged surfaces available. This is even more important in the next stage.

Samples for Chemical Analysis

Before any sample cutting or nondestructive testing is performed, it is important that corrosion products, paint, coating samples, deposits, or representative surfaces with in situ films be removed for further examination. If a corrosion product film is not removed before further testing then the fluids from NDE (dye, ultrasonic couplant, etc.) or heat from cutting the sample may interfere with or change the composition. This also holds true for many paints and coatings.

Nondestructive Evaluation (NDE)

This is a large field unto itself and many texts are available to aid in deciding which NDE method is appropriate for a failure analysis. The most common testing methods are liquid (dye) penetrant, magnetic particle, electromagnetic, radiographic, and ultrasonic. All of these methods have advantages and disadvantages over the others and it is often necessary to use more than one method to fully characterize defects and imperfections in a component. However, it is often advisable to use NDE prior to destructive testing of components to ensure a complete picture of the failure is obtained prior to cutting samples for additional analysis. Moreover, it gives the failure analyst the opportunity to select the best samples for further analysis.

Macroscopic Examination

Macroscopic examination, which includes the use of hand lenses and low power (~25X) microscopes, can provide a significant amount of information, yet is often bypassed in favor of high magnification microscopic techniques. In fact, macroscopic and microscopic examination are complementary and both are necessary. For example, the determination of the fracture origin in a long running brittle fracture of a pipeline can easily be traced by following the chevron pattern on the fracture surface back to the origin. Figure 7–1 is an illustration of

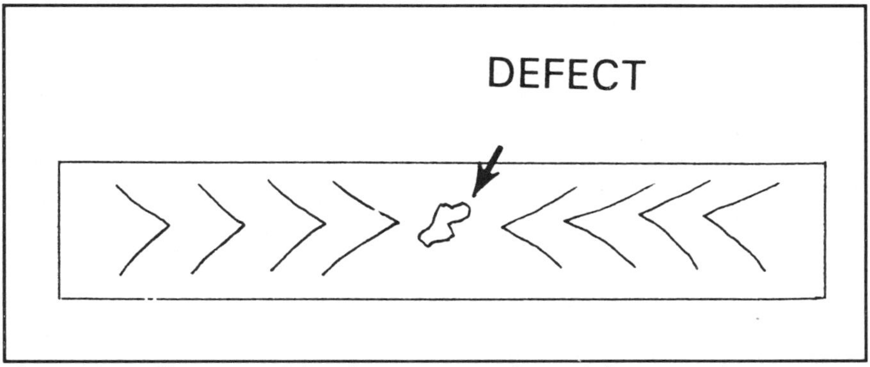

Figure 7–1 Illustration of the chevron pattern formed during crack propagation and the location of the crack origin.

how chevrons on the fracture point back to the origin. This can save significant time in examining only the origin for the cause of failure rather than the large amount of crack propagation surfaces that are actually secondary to the failure. Figure 2–9 in Chapter Two is a good example of macroscopic examination. The fracture origin in fatigue of a sucker rod is near the flat portion of the failure while the final tensile overload shows a 45° shear lip. Figure 7–2 is an example of environmental cracking of an S–125 coupling. The fan-shaped pattern emanates from a single point; the origin of the fracture. Thus, the failure analyst can concentrate the analysis on the fracture initiation area and not other areas of the fracture that were not pertinent to the cause.

If the macroscopic features of the fracture display necking, then the failure was probably ductile and further testing will often take a different course than if the features are more brittle in appearance. Brittle fractures are usually flat in appearance. However, care must be taken in using generalities. For example, Figure 7–3 shows a schematic of ductile and brittle macroscopic fracture appearance under tensile loading and the reversal of this pattern under torsional loading. Thus it is important to know the stress history as well.

Sample Removal and Preservation

After all visual and macroscopic work has been completed, samples for mechanical testing, chemical analysis, metallography, and electron microscopy can be removed. All remaining pieces should be carefully preserved and stored for future work. Frequently in failure analysis, after results of the above mentioned

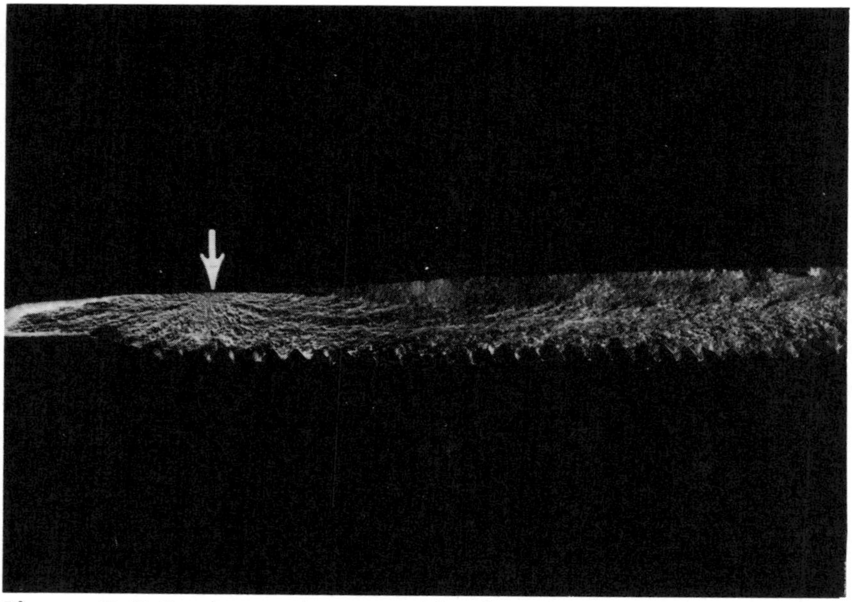

Figure 7–2 Crack origin (arrow) and fan-shaped pattern emanating from it on an S-125 coupling that failed from environmental cracking. Shown at original size.

destructive results are obtained, either further confirmation testing is necessary or the results send the investigation into another direction that may require additional, albeit different, tests to confirm different mechanism.

Mechanical Testing

Mechanical testing can be as simple as a standard tensile test or hardness test or may involve sophisticated fracture mechanics testing. Depending on the original specifications for manufacturing or operation, impact testing, creep testing, etc., may be necessary to evaluate the failed part in comparison to the original requirements.

Mechanical testing for failure analysis is probably misapplied as often as it is correctly applied. For instance, tensile testing a component that has failed from hydrogen damage several weeks after the failure may show no effect of hydrogen because the hydrogen has had sufficient time to diffuse back out, returning the alloy to its original ductile condition. Similarly, high strain-rate impact testing

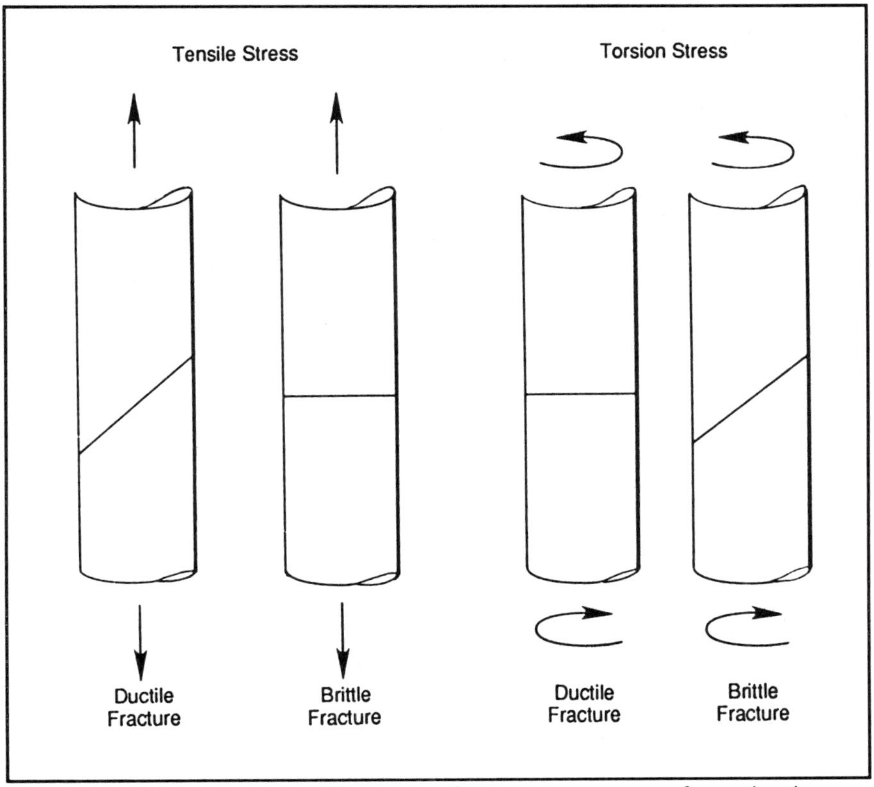

Figure 7–3 Macroscopic difference in fracture appearance for a ductile fracture and a brittle fracture under a tensile load and the same appearance under a torsional load.

(Charpy V-notch) is not sensitive to hydrogen damaged steels, unless irreversible damage has occurred. Thus it is crucial that the analyst utilize the correct tests to evaluate certain properties, otherwise the data and the analysis will be misleading.

Microscopic Examination

Both metallography and electron microscopy have been included in this section. While scanning electron microscopy (SEM) and transmission electron microscopy (TEM) can both be valuable tools in failure analysis, the latter is

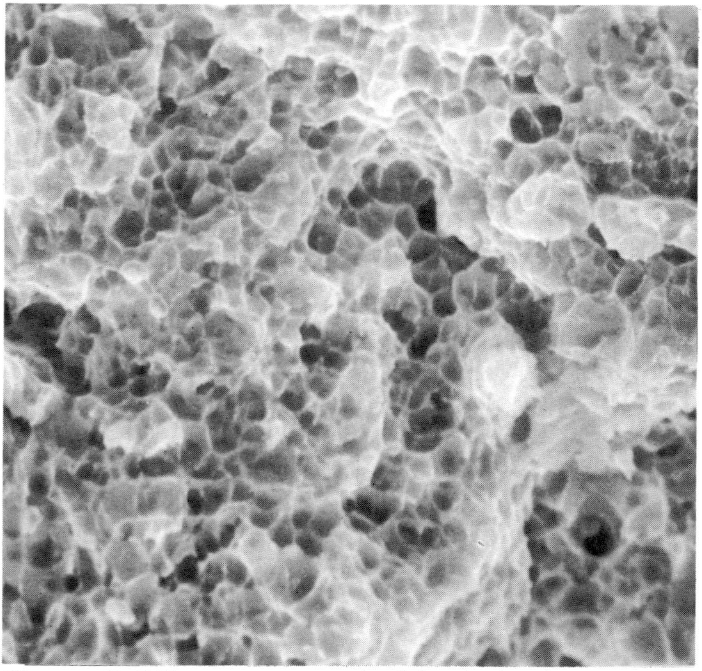

Figure 7–4 Electron micrograph of ductile dimple rupture, also referred to as microvoid coalescence.

infrequently used since it primarily views the internal structure of metals while the former is for surface examination.

Metallography has been extensively used in this book to provide examples of materials and failures in the oil and gas industry. It is a valuable and indispensable tool to evaluate the microstructure of the material and investigate various types of failures.

Microscopic examination using SEM has been less often presented in this book but also is extensively used in failure analysis. At low magnification, 1 5–50X, it can be useful for determining the fracture origin or examining the morphology of cracks and corrosion pits. At higher magnification, SEM provides a key to the nature of a cracking process and if changes occurred from initiation to propagation. There are several standard fracture morphologies or modes that are frequently observed and are useful to the failure analyst.

Figure 7–4 to 7–7 show examples of some of these modes. Dimple rupture, also referred to as microvoid coalescence, which is a ductile failure mode is shown in Figure 7–4. Transgranular tearing, another ductile fracture mode is not

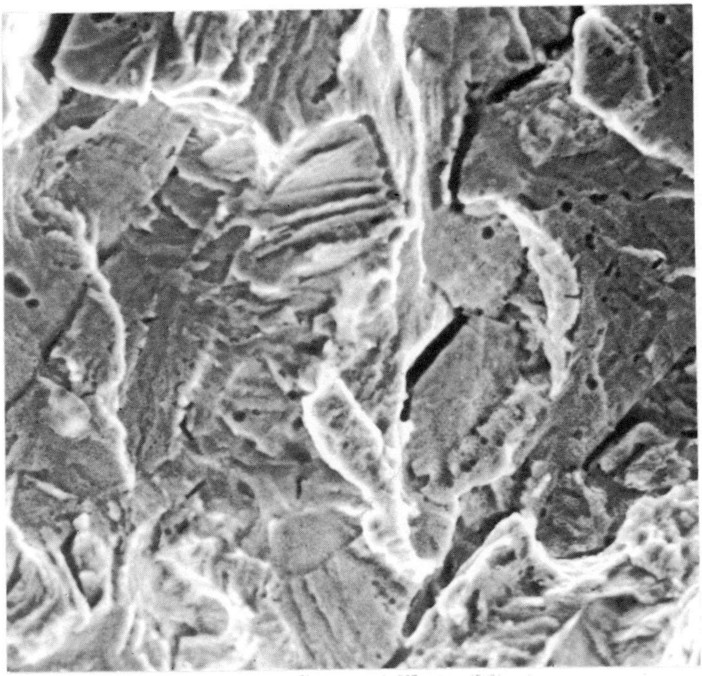

Figure 7–5 Electron micrograph of a quasi-cleavage fracture in steel.

shown but appears feathery. Quasi-cleavage, which is a brittle transgranular fracture mode, is illustrated in Figure 7–5. Cleavage that is also a brittle transgranular fracture mode but follows more crystallographic planes than quasi-cleavage is shown in Figure 7–6. Intergranular brittle fracture is shown in Figure 7-7.

These various fracture modes and combinations of each aid in determining the cause of a failure.

Chemical Analysis

During this stage, bulk chemical analysis of the components can provide information on the composition of the alloys and whether they meet the original requirements. Moreover, coatings and paints can be tested for composition.

In combination with SEM work, the surface can be analyzed by either wavelength or energy dispersive spectroscopy (WDS or EDS) for the presence of

Figure 7–6 Electron micrograph of cleavage fracture.

different elements both in the alloy and on the surface. Generally, only a semiquantitative analysis can be performed and the composition of *compounds* cannot be determined. Thus compounds must be inferred by this analysis.

Newer surface science techniques can be used to gain a better understanding of surface composition than can be obtained by WDS or EDS. Moreover, they are much more sensitive to low concentrations than WDS and EDS. Two of the most useful are Auger Electron Spectroscopy (AES) and X-ray Photoelectron Spectroscopy (XPS) or as it is also known, Electron Spectroscopy for Chemical Analysis (ESCA). Both of these surface analysis techniques can measure elemental composition as low as one atom layer thick on the sample and can detect all elements except hydrogen. XPS can also provide information on the bonding states of different elements so that compounds can be described. There are numerous other surface science techniques, but these two have already proven to be quite useful in failure analysis in the oil and gas industry.

Bulk crystallized compounds removed from the surface of a sample can be analyzed by X-Ray diffraction (XRD) for compound composition, and those compounds not well-crystallized can be analyzed for elemental composition by X-ray fluorescence (XRF).

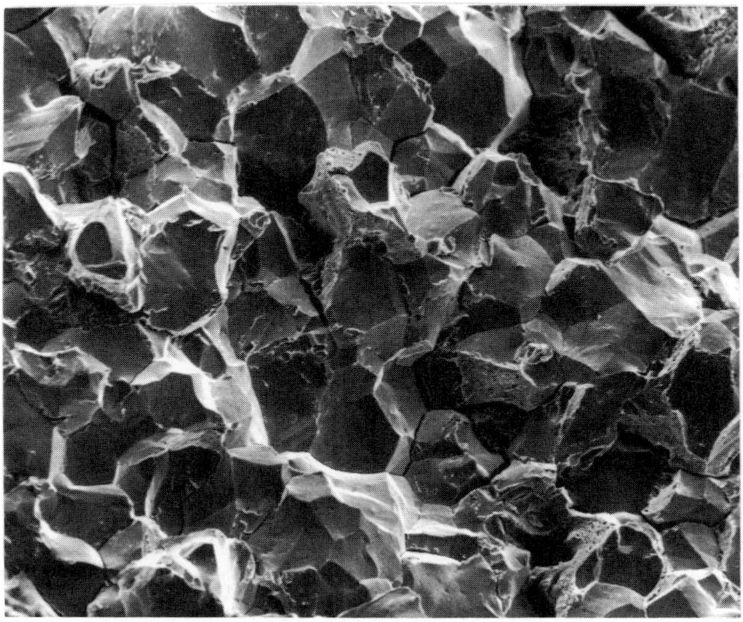

Figure 7–7 Electron micrograph of intergranular fracture.

Stress and Fracture Mechanics Analysis

In some cases it is necessary or desirable to perform a stress analysis of the component and/or analyze the fracture mechanics involved. This may take the form of simple calculations or the requirements of detailed computer analysis using finite element techniques. At this time, more mechanical testing may be required to augment the mathematical analysis.

Review of Data, Formulation of Failure Scenario Report

Once the data are accumulated and analyzed, the number of possible causes for a failure are considerably reduced, sometimes to just one reasonable cause but at other times to several possible causes. Often more evaluation and investigation is necessary, either through additional testing or review of the technical literature to eliminate those possibilities for failure that are least likely and maintain the one or two that are the most reasonable in light of the evidence. In many failure analyses, the cause of failure is finally determined through deductive rather than inductive reasoning. It is at this stage that many potentially good and correct failure analyses fall short by reaching an incorrect conclusion. During this stage, the conclusions and purported cause of failure must be substantiated by the data and not merely reflect the wishes or bias of parties involved in the investigation.

Once the analysis is complete, it is prudent to produce a written report for the benefit of those parties involved in the particular failure but also to provide a record for the future that can be an aid in eliminating or reducing such failures or improving the design and operation of a component.

References

ASM International, Metal Handbook, *Failure Analysis and Prevention,* Vol. 10, (1975).

ASNT, Nondestructive Testing Handbook, Second Edition, Vol. 1–5, (1975–1987).

INDEX

I

L

M

N

O

P

Q

S

T

U

W

Y